A Physicochemical Perspective On Evolutionarily Related Repeated, Interspaced, Palindromic Sequence-Associated Effectors For Editing Genetic Material

Lucia Leon

A Physicochemical Perspective On Evolutionarily Related Repeated, Interspaced, Palindromic Sequence-Associated Effectors For Editing Genetic Material

First Edition March 2025

Written by Lucia Leon

Table of Contents

A Biophysical insight into orthologous CRISPR-associated (Cas) proteinfor genome editing

Chapter 1
Review of Literature

1.1. Structure and mechanism of Cas9 protein and its orthologue: An Overview

The CRISPR/Cas9 system is a naturally occurring, adaptive heritable immune system which defends invading phage attack and present in the whooping numbers of prokaryotes.This has been repurposed as a powerful tool which is guided by RNA and targets DNA for genome engineering which works by making site-specific DNA break and its subsequent repair by cellular machinery. The simplicity and multiplexity of the CRISPR system make it more sought after than other older counterparts viz. ZFNs and TALENs, and this has led to harnessing of CRISPR in gene therapy programmable gene regulation, cellular imagings, synthetic biology and nucleic acid detection among others. The CRISPR genome editing system includes an endonuclease, Cas9 derived from type II CRISPR-Cas system and a chimeric single-guide RNA (sgRNA). Cas9 and sgRNA forms ribonucleoprotein (RNP) binary complex and upon recognition of a protospacer adjacent motif (PAM) it binds to the substrate DNA, and forms ternary complex by unwinding the PAM upstream region with subsequent formation of R-loop. Successful R-loop formation displaces PAM-containing DNA strand and results in formation of a heteroduplex only when the target DNA strand finds complementary to sgRNA sequence. It is worth mentioning Cas9 remains inactive inthe absence of sgRNA and sgRNA binding induces structural changes in Cas9 from apo toactive form.

Cas9 is an endonuclease (~158 kDa) with a bi-lobed architecture consisting of α-helical recognition (REC) and nuclease (NUC) lobes. The REC lobe comprises of mainly three regions namely bridge helix, REC1 domain, and REC2 domain. REC1 adopts an elongated, α-helical structure containing α helices and β sheets, whereas REC2 adopts a helix containing bundle-like structure. The NUC lobe contains the

PAM-interacting (PI) domain,which forms an elongated structure comprising of α helices and β sheets. NUC lobe is composed of two separate domains, RUVC and HNH alongwith a C-terminal domain (CTD) which interacts with the PAM duplex DNA. Though structural studies of Cas9 using X-ray crystallography and cryo-EM unveil domain architecture of Cas9, single-molecule Förster resonance energy transfer (sm-FRET) techniques capture conformational dynamicsand kinetics of Cas9 during binding and cleavage of DNA substrate, thermodynamic details of Cas9 and sgRNA binding followed by target DNA interrogation by Cas9 RNP is unclear. Here, we first report formation of Cas9 ternary complex is enthalpically deficit denoting the large positive change in enthalpy.

Despite significant strides in mechanistic understanding of Cas9 function, few fundamentals in terms of thermal stability of Cas9, its salt-dependent structural changes, Cas9 activity (binding and cleavage) as a function of salt concentration and temperature isutterly lacking. Previous research revealed that culture temperature has a significant impact on the effectiveness of genome editing. Cells were cultured in transient hypothermia conditions (30°C), which significantly improved zinc finger nuclease (ZFN) and transcription activator-like effector nuclease (TALEN) mediated editing efficiency.

1.1.1. Brief history of CRISPR-Cas9 system

In 2002, Ruud Jansen and colleagues coined the term CRISPR (clustered regularly interspaced short palindromic repeat DNA sequences) (Jansen et al. 2002). The discoveryof key CRISPR repeat and spacer regions came as a result of the sequencing and analysisof an increased number of bacterial genomes. CRISPR sequences are found in nearly halfof all sequenced genomic DNA of bacteria and 90 percent of archaea (F. J. Mojica et al. 2000). These CRISPR elements are identified alongside a number of well-preserved genes which are later designated as CRISPR-associated

(Cas) genes (Jansen et al. 2002). Furthermore, the non-repeating spacer nucleotide sequences of the CRISPR element havebeen recognized as belonging to viruses (Bacteriophages) and other foreign genetic elements (DNA) (F. J. Mojica et al. 2000; Pourcel, Salvignol, and Vergnaud 2005; Bolotin et al. 2005). As a result of these findings, many researchers became eager to investigate the functional importance and technical elements of these CRISPR sequences. Several investigations have determined that it is a bacteria's adaptive immune defence against bacteriophages and other foreign genetic DNA material. Barrangou and colleagues published significant experimental evidence for the likely function of CRISPRsystems in 2007, claiming that bacteria insert new spacers formed from phage nucleotide sequences into their genome following a viral assault. Finally, the CRISPR spacer sequences acquired from phages to confer protection against certain phages dictate thetargeting specificity of Cas enzymes (Barrangou et al. 2007). Since then, the CRISPR-based defence in bacteria and its flexibility for gene editing in mammalian cells have been the subject of several investigations.

Brouns and colleagues discovered that Cas enzyme activity is directed by short CRISPR RNAs (crRNA) recognized from CRISPR elements, which include spacer sequences anda part of repeat motifs (Brouns et al. 2008). These crRNAs also block phage and horizontal DNA transfer into the host bacteria, according to further studies (Deveau et al. 2008). According to additional research, the generated spacer sequences are nearly identical at protospacer-adjacent motifs (PAMs), a critical region for the CRISPR system's operation. It was also discovered that only Cas9, out of the several Cas proteinsknown, possessed DNA catalytic activity (Garneau et al. 2010). Cas9 is a composite enzyme with two nuclease domains: an HNH domain for cleaving the target DNA strand (specified as complementary to a certain spacer/crRNA sequence) and a RuvC-like motiffor cutting the non-target strand (Jinek et al. 2012; Sapranauskas et al. 2011; Gasiunas et al. 2012; Makarova et al.

2006). Moreover, biochemical analysis of the different CRISPR Cas system components showed that the endogenous CRISPR system needs two short RNAs: the matured crRNA and a Trans activating crRNA (tracrRNA). The crRNA is made up of a guide sequence and a complementary repeat element that pairs with the tracrRNA. The tracrRNA is involved in a number of events, including the maturation of crRNA from a pre-mature crRNA transcript and the binding of Cas9 to theguide RNA. The guide RNA, which is made up of crRNA and tracrRNA, forms a complex with the Cas9 protein in order to cleave the target DNA by creating double strand breaks (DSBs) at the target region (Deltcheva et al. 2011). Jinek and colleagues later demonstrated that combining tracrRNA and crRNA into a single full length intact RNA called single guide RNA (sgRNA) may similarly lead CRISPRCas9 to break DNA at a specific spot (Jinek et al. 2012). More and more studies come up with groundbreaking publications that modified and repurposed the CRISPR Cas9 system for *in vivo* genetic modification in eukaryotic cells (Mali, Yang, et al. 2013; Jinek et al. 2012;Cong et al. 2013).

1.1.2. CRISPR-Cas9 systems' classification:

In 2015, Makarova and colleagues classified them into 2 classes, 6 types and 33 subtypes based on the on the effector module genes, i.e., whether the systems contain numerous variants of a multisubunit complex (Cascade) complex, like the Csm and Cmr complex or a single protein like Cas9. The most comprehensive level of classification for CRISPR–Cas systems, with 'class 1' and 'class 2' divisions.

Class 1 systems have multisubunit crRNA–effector complexes, whereas class 2 systemshave a single protein, such as Cas9, that performs all of the effector complex's tasks.

Further, they classify these 2 classes into 5 types with class 1 having type I, type III andtype IV and class 2 with type II and type V (Makarova and Koonin 2015). They further classified into 33 subtypes.

Now since we have come across with a lot more discoveries of different Cas proteins then they are further classified into different classes and types and subtypes. The new classification was done based on the previously used approach like the presence of trademark Cas proteins, phylogenies of the most evolutionarily conserved Cas proteins,sequence homology between Cas proteins, and gene arsenal and genomic locus organization comparisons. Here they classify them into 2 classes as done previously, but the types have been raised from five to six, and an additional 17 subtypes have beenadded, bringing the total number of subtypes to 3 **(Fig. 1.1)**. New subtypes were assigned to loci which did not match the criterion for incorporation in any of the previously defined subtypes.

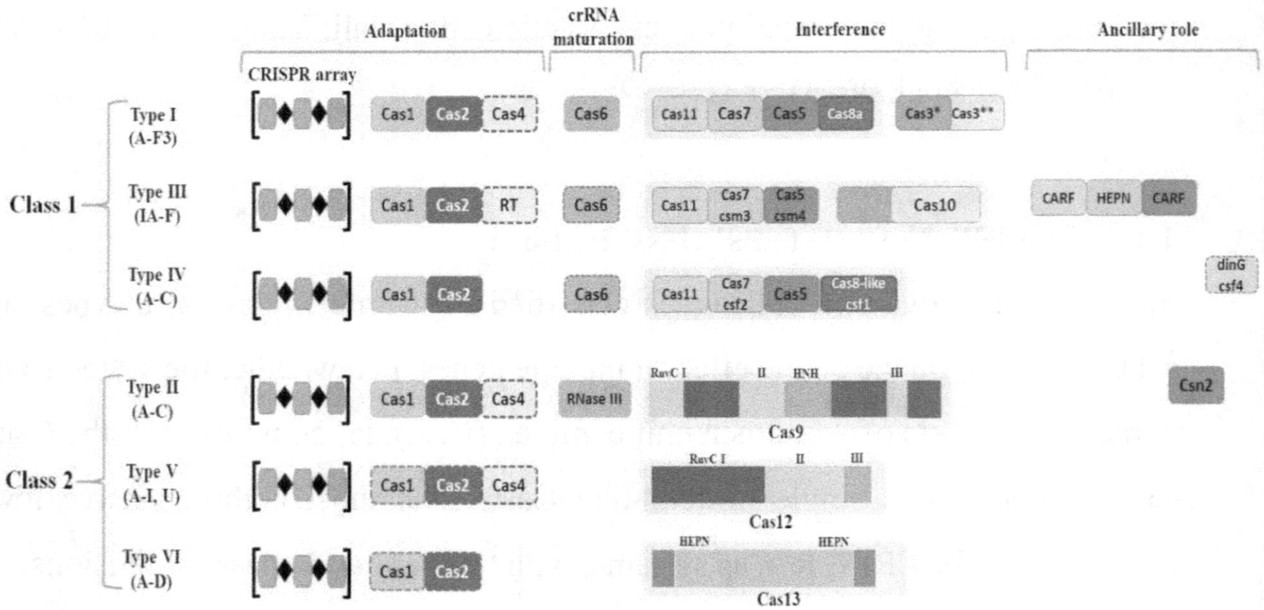

Figure 1.1: Classification of CRISPR Cas system: Schematic is adapted from the Makarova et al., 2019, showing the general architecture of functional modules for 2 classesand 6 types of CRISPR–Cas systems.

1.1.3. Mechanism of action of CRISPR-Cas9 system

CRISPR-Cas has appeared as a powerful tool in gene therapy with potentially far-reaching effects (Hille et al. 2018; Hsu, Lander, and Zhang 2014; Komor, Badran, and Liu 2017). The CRISPR-Cas9 method has organized the science of gene editing. Becauseof its protein-based gene-editing ancestors, the ZFNs and TALENs, it is a simple RNA- led DNA-binding tool that uses a bacterial endonuclease and is easy to produce and precise in function. A 20-nucleotide long CRISPR RNA (crRNA) with a complementary sequence to a specific DNA site guides this configurable and highly multiplexable platform, which is preceded by an approximately 80-nucleotide stretch of transactivating RNA (tracrRNA), which can attach and employ the Cas9 protein to its target location

(Garneau et al. 2010). The bacterial endogenous CRISPR-Cas9-based defence response has three key phases **(Fig. 1.2)**. First, the adaptation phase occurs when exogenous genetic components enter the system and activate a mechanism that recognizes and processes the protospacer. The protospacer is then inserted as a new spacer in the CRISPR repeat regions of the host chromosome. These spacers are stored as a genetic memory for the following stages. The second step is the expression phase, after reinfection with phage or foreign genomic components, the machinery for Cas9 gene expression and CRISPR element get triggered in the expression phase. Beginning with the leader sequence, pre-crRNA transcription generates many repetitive and spacer sections. These regions are then modified into mature CRISPR RNAs (crRNAs) by Cas proteins and along with that the Cas9 protein is directed towards its target sequence. Theinterference phase which is the third and final step, where the processed guide RNA andthe Cas9 protein (Type II system) form a complex that identifies and eliminates foreign DNA fragments inside the cell (Makarova et al. 2011; van der Oost et al. 2009).

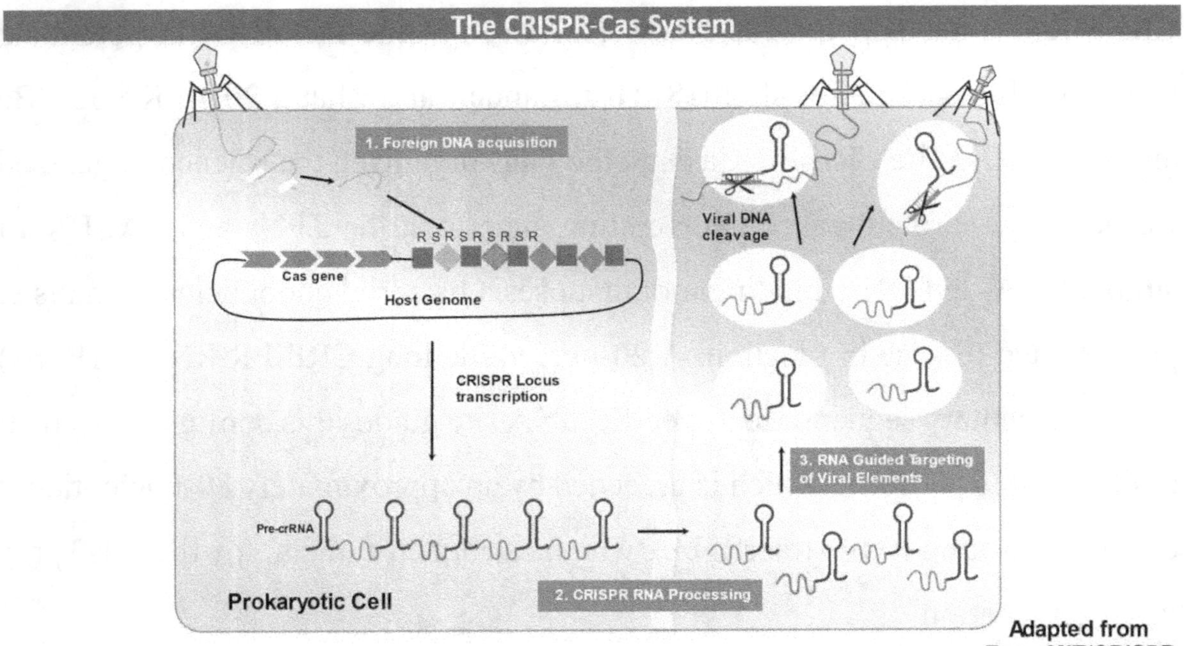

Adapted from Team:MIT/CRISPR

Figure 1.2: The CRISPR-Cas9 mediated adaptive immunity in bacteria.

A schematic showing the different stages of bacterial CRISPR-Cas9 based adaptive immunity against the bacteriophage infection.

1.1.4. Orthologues of CRISPR-Cas9

The discovery of the CRISPR-Cas system has made it feasible to do reliable and affordable genome editing in a wide range of cells and model species. *Streptococcus pyogenes* Cas9 provides high levels of activity but has been associated with shortcomingslike off-targeting. The shortcomings are as follows - firstly, at target sequence or genomic sequences the Cas9 enzyme showed the mutagenesis; secondly, the target zone may be constrained by the necessity of a PAM motif nearby the chosen target site; Lastly, the efficient delivery of Cas9 and the other components of genomic engineering tool is challenged by the sizes of Cas9. To facilitate its efficient delivery,

Cas9 has also been shortened and engineered into component.The improvements in efficiency of delivery, accuracy and target range will ease the therapeutic application of site-specific CRISPR nucleases. For this thesis, we are interestedin following Cas9 orthologues only- SpCas9, FnCas9, CjCas9 and SaCas9.

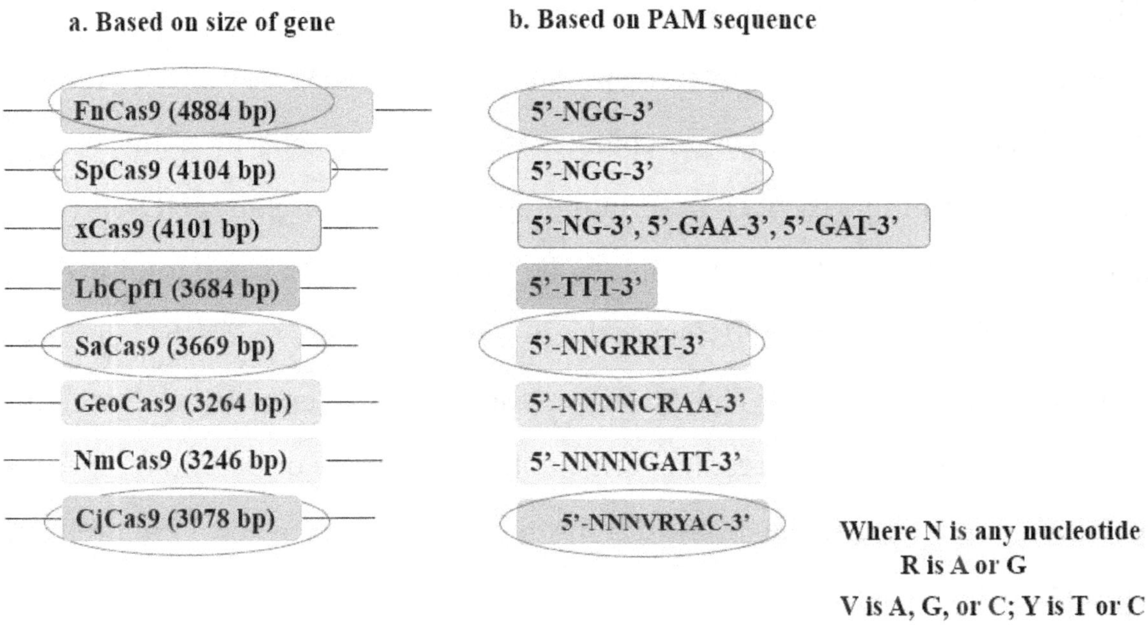

Figure 1.3: Choosing the orthologous proteins of CRISPR-Cas system based on thesize of gene and PAM sequence

1.1.5. CRISPR-Cas9 genome editing

The RNA-guided Cas9 nuclease looks for and recognizes a conserved nucleotide sequence known as the protospacer-adjacent motif (PAM), a nucleotide sequence unique to a Cas9 ortholog, present adjust to the DNA target region (F. J. M. Mojica et al. 2009; Horvath et al. 2008; Deveau et al. 2008). Upon binding to the PAM, Cas9 melts a few bases of local double stranded DNA and examines adjacent DNA sequences for base- pairing complementarity to the guide RNA after binding to the PAM (Sternberg, Haurwitz, and Doudna 2012). If the first few bases of the guide RNA and the target DNA strand have complementarity, the guide RNA strand invasion leads to more local DNA unwinding and base pairing, forming an R-loop (Szczelkun et al. 2014). The R-loop formation causes conformational changes that activates the endonuclease domains to accurately cleave each DNA strand via the RuvC and HNH domains, resulting in a double-strand DNA break (DSB) a few bases away from the PAM (Garneau et al. 2010).

Furthermore, The cell's own repair system then steps in to mend the double strand breaks(DSBs), causing nucleotide changes in the target DNA location through non-homologous joining (NHEJ) and homology directed repair (HDR) (Hille and Charpentier 2016; van der Oost et al. 2014). Non-homologous end joining (NHEJ) or Homology Directed Repair(HDR) are the two processes by which the cell recognizes and repairs DNA damage. The NHEJ method uses an error-prone mechanism to repair a DNA break, which results in nucleotide insertions or deletions (indels) at the break site. As a result, the frame for protein coding sequence is altered, perhaps resulting in the deletion of the desired protein.Using CRISPR and the biological NHEJ repair system to knock off a gene is straightforward, rapid, and inexpensive. On the other hand, the HDR process uses a very precise mechanism to repair the break using a template DNA strand in such a way that an identical replica of the original DNA is generated with no loss of information.

Unfortunately, such an approach is not very efficient, since it necessitates the coordinatedaction of a huge number of recombination proteins in the presence of a template.

Nonetheless, it raises the prospect of an exogenous DNA segment with homologous armsBeing incorporated into the DNA break site. As a result, CRISPR creates a pathway for foreign DNA to be inserted at a specific spot, allowing for precise DNA editing with highefficiency.

1.1.6. Applications of CRISPR-Cas9 system

The features of CRISPR–Cas9, when combined with knowledge from other genome-editing techniques, enabled quick acceptance for genome engineering applications in various aspects of biology (Barrangou and Marraffini 2014). The flexibility and efficacy it gives results in its utility in practically all cell types, including medically important cells and animals, agriculture and livestockrelated plants, and animals and model organisms extensively recognized by researchers (Doudna and Charpentier 2014; Ledford2015). CRISPR–Cas9 has been used to build better genome-editing techniques in animals, plants, and microbes. Its ability to make DNA double break followed by endogenous repair method's feature can also be employed to treat various genetic diseases like sickle cell anaemia (SCA). Using CRISPR-Cas9 and the HDR pathway machinery, researchers were able to successfully substitute the wild type HbB gene in CD34+ HSCs from SCA patients (Wilkinson et al. 2021; H. Park et al. 2019; Dever et al. 2016). Using the CRISPR-Cas9 technology, genome editing has been recognized to produce accurate heritable changes in the genome of organisms. Despite the fact that the widely recognized Streptococcus pyogenes Cas9 (SpCas9) has been effectively exploited for diverse genome-editing activities across various platforms, worries about possible off-targeting at several loci across the genome remain.

1.2. Genome editing nucleases

In the 1970s, the discovery of restriction enzymes, bacterial nuclease proteins that help protect bacteria from phage infection and other foreign DNA, was a turning point in recombinant DNA technology (Danna and Nathans 1971; Kelly and Smith 1970; H. O.Smith and Wilcox 1970). For the first time, scientists have managed to tweak DNA in a test tube. This, together with revolutionary work on homologous recombination to integrate foreign DNA into the genome, ushered in a new era of genome editing (Smithies et al. 1985). Further, the confirmation of the hypothesis introducing precise double strand breaks (DSBs) improves the success of targeted gene integration, many researchers began looking for nucleases that could recognize and cut DNA at specific target locations. This concept ushered in a new phase of genome editing and targeting.

1.2.1 Homing endonucleases

Homing endonucleases, also known as Meganucleases, are naturally occurring endonuclease enzymes that have a high selectivity for cutting DNA at specific loci. They've been discovered in nearly every microorganism, as well as mitochondria and chloroplasts (Michel, Grandchamp, and Dujon 1980; Lazowska, Jacq, and Slonimski 1980; Dujon 2005). They were found to include parasite DNA elements, such as inteins or group 1 self-splicing introns (protein introns). They recognize DNA cleavage sites of upto 30 nucleotides and have been reconstructed for different targets andrecognize DNA cleavage sites of up to 30 nucleotides (P. P. Silva et al. 2011; J. C. Silva et al. 2006). They have a high specificity, are tiny, and produce 3' overhangs (advantageous for Homology directed repair (HDR) events). Mammalian genomes have been targeted using meganucleases like I-Sce I and I-Cre I (Plessis et al. 1992; Cohen- Tannoudji et al. 1998). Even though hundreds of other naturally occurring meganucleaseshave been discovered, the chances of identifying a meganuclease for a specific loci are still rare.

1.2.2 Zinc-Finger Nucleases (ZFN)

Due to the limitations of Meganuclease enzymes, researchers seek for new techniques to directly target the genome and make double strand breaks. Zinc finger nucleases (ZFNs) were discovered and engineered to target specific sites, making them a viable alternative to meganucleases (Y. G. Kim, Cha, and Chandrasegaran 1996; Klug and Rhodes 1987). Zinc finger protein (ZFP), a small protein motif regulated by the zinc ion, found to be oneof the most common proteins in eukaryotic genomes. The zinc fingers are able to identify and adhere to a 3-bp DNA sequence (Klug and Rhodes 1987). Zinc-finger nucleases are a hybrid of Zinc-finger and FokI nucleases that emerged from structural research on zinc-finger proteins and Meganucleases. The Fok I nuclease is made up of two domains: one for specific DNA binding and the other for non-specific DNA cleavage at a distance. The DNA binding domain has a limited application range in terms of genome editing because it can only bind to a certain sequence. The combination of FoKI's non-specific nuclease domains with a zinc-finger motif as a DNA binding domain improves the activity and specificity of nuclease-based double strand breaks and downstream processes. Since FokI works as ahomodimer, dimerization requires two zinc-finger modules linked to it to be close together, resulting in a double strand break (DSB) at a target site (Y. G. Kim, Cha, and Chandrasegaran 1996). Each zinc finger module binds to three bases, and combinations of 6-7 of these fingers can recognize 18-21 nucleotides, yielding a huge number of sequence combinations for almost any place in the genome **(Fig. 1.4)**. ZFN's design for diverse locations has been improved, allowing it to access and cut any site (Lombardo et al. 2007; Urnov et al. 2005; Miller et al. 2007). Despite their potential, the time it takes tocustom-engineer ZFNs with high specificity and affinity for DNA target sites has impeded their widespread use (Ramirez et al. 2008). Furthermore, ZFN's efficacy is limited by the quantity and placement of viable target sites within a genome.

1.2.3. Transcription activator-like effector nuclease (TALEN)

TALE proteins are a type of bacterial plant pathogen protein produced by the genus Xanthomonas. They are responsible for transporting proteins to plant cells during infection (Bogdanove et al. 2010). They bind related protein DNA sequences in plant cells and induce gene expression via transcription. The chimeric integration of the

Fok I DNA cleavage domain with a TALE module acts as a valuable tunable nuclease with unique specificities to create double-strand breaks at specific genomic loci, analogous to ZFNs (Doyle et al. 2011; Christian et al. 2010; F. Zhang et al. 2011). In principle, the Transcription activator-like effector nucleases (TALENs), like ZFNs and homing endonucleases, are tailored for each DNA target sequence in principle **(Fig. 1.4)**. In theircore region, these TALE proteins have a variable amount of tandem DNA repetitions of amino acid sequences, which aids in DNA binding. Each repeat's amino acid sequence isalmost identical except for two adjacent amino acids at positions 12 and 13, (the repeat variable di-residue or RVD). Variations in the number of repeats (13–33) and the RVD makeup are used to identify the DNA base pair. The specificity of TALE proteins is based on a simple coding system in which the RVD of a repetition specifies the one baseof the DNA target region. As a result, the DNA interaction site in a single TALE protein is defined by a sequential arrangement of repeats with various RVDs (Christian et al. 2010; F. Zhang et al. 2011). TALENs have an advantage over ZFNs in that each module recognized only one nucleotide in the targeted DNA sequence, resulting in less interference between modules and hence a more specific recognition (Boch et al. 2009). They've shown to be a lot easier to design and construct in a short amount of time, and they can be quickly expanded to any desired length (Cermak et al. 2011). In comparison to ZFNs, whole-genome sequencing investigations of TALEN edited cells revealed a fairly low overall off-target incidence rate (C. Smith et al. 2014). A clear disadvantage ofTALENs over ZFNs in terms of size, which is three times that of ZFNs. Furthermore, TALENs have a highly repetitive sequence that impairs their ability for viral vectors to package and distribute them (Holkers et al. 2013).

1.2.4. Clustered Regularly Interspaced Short Palindromic Repeats (CRISPR) system

CRISPR stands for clustered regularly interspaced short palindromic repeat DNA sequences, which are involved in the immune systems of bacteria and archaea. It operatesas a protein-RNA complex to make a DNA double stranded break (DSB) at a specific location. The RNA works as a guide to direct protein to the target DNA, while the protein(Cas protein) acts as a nuclease enzyme to cleave the DNA. Despite the promise, the intricacy of cloning and protein engineering has delayed scientific acceptance of ZFNs and TALENs. Unlike ZFNs and TALENs, which require the design and engineering of two protein nucleases for DNA recognition, the CRISPR-Cas system in genome engineering relies on a short RNA for DNA recognition (Deveau et al. 2008; Barrangou et al. 2007).

Compared to building two nucleases proteins to recognize two locations close of a genomic target sequence as in the case of ZFNs or TALENs, designing short RNA (20-30nt) is considerably easier and faster (Gasiunas et al. 2012; Jinek et al. 2012) **(Fig.1.4)**.

The fundamental advantage of the bacterium type II CRISPR-Cas system is the minimum requirement for customisable DNA interference: an endonuclease, Cas9, controlled by a tinkerable dual-RNA structure. Other benefits of the CRISPR-Cas9 method include its efficiency, reliance on a consistent Cas9 protein, capacity to address numerous targets simultaneously with various guide RNAs, and low cost, making it accessible to small labs(Cong et al. 2013). Furthermore, it is also relatively easier to use and more flexible.

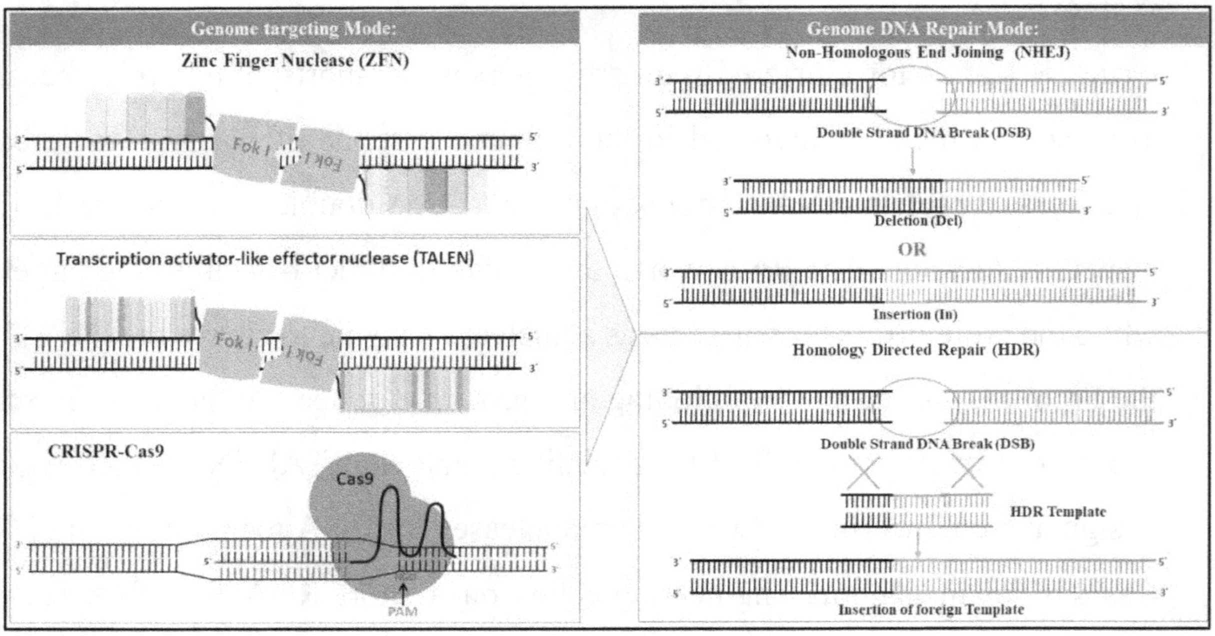

Figure 1.4: Different platforms for Genome editing and repair mechanisms for endogenous double strand DNA breaks (DSB): Schematic showing the Genome editingnucleases (ZFNs, TALENs and CRISPR-Cas9) introducing DSBs at targeted sites (Left). The endogenous repair machinery repairing the DSBs either by non-homologous end joining (NHEJ), leading to form insertion or deletion in the target site, or via Homology directed repair (HDR) pathway in the presence of donor template, leading to precisely addor correct the target sequence (Right).

1.3. Biophysical assays used in literature for CRISPR Cas9 system

CRISPR (clustered regularly interspaced short palindromic repeats) - associated (Cas) proteins and CRISPR RNAs (crRNAs) are combined in the bacterial adaptive and heritable immune system to bind foreign nucleic acids. A type II CRISPR-Cas effector complex called Cas9 binds to a single guide RNA that base pairs with the target strand of double-stranded DNA. This displacement of the non-target strand results in the formation of a R loop, where the HNH and RuvC nuclease domains can cleave the target strand and non-target strand of double-stranded DNA, respectively. Biophysical studies of RNA guided Cas9 complex-DNA interactions is of fundamental importance for the developmentof CRISPR/Cas9 binding mechanism. Though many structural, biochemical and next generation sequencing studies have already been done, that shed light on the mechanism ofCas9 cleavage and specificity of the Cas proteins towards DNA targets, a unifying biophysical study of Cas9 is yet to emerge which will represent the binding affinity and kinetics for each step of the entire mechanism providing information like dissociation equilibrium constant and rate constant. This section of the thesis seeks to present a comprehensive structural and mechanistic understanding of the Cas9-mediated RNA-guided DNA targeting, cleavage, and repair process of genome editing in relation to kinetics from the viewpoint of biophysics. Here, molecular mechanistic insights from biophysical, structural, and biochemical studies offer a framework for rational engineering aimed at varying catalytic function, PAM needs, guide RNA specificity, and lowering off-target activity for the development of Cas9-based therapies by genetic editing.

The CRISPR (clustered regularly interspaced short palindromic repeats)-associated cas9 protein system, a bacterial adaptive defense mechanism against foreign genetic element transfer such as those present within plasmids and phages is generally found in approximately 50% of sequenced bacterial genomes and almost 90% of sequenced archaea. For genome editing, epigenetic regulation, transcriptional disruption, and genome imaging, this naturally occurring system has been used as a potent RNA-guided DNA targeting tool. Through the use of genome engineering, it is now possible to alter any genomic sequence that is specified by a short stretch of guide RNA. This technology also enables the correction of disease-causing mutations, the inactivation of activated oncogenes, and the activation of dormant cancer suppressor genes. CRISPR-based genome-wide screens can be used to find novel tumour suppressors or other genes that are disease- or drug-resistance targets. Sickle cell anaemia, cystic fibrosis, viral infections, immunological problems, cardiovascular diseases, and Duchenne are just a few of the genetic conditions that CRISPR-Cas9 genome engineering has the potential to treat or possibly cure.

Despite its advantage and great promise, there are some difficulties in using CRISPR–Cas9for therapeutic purposes. The specificity for the gene targeted with Cas9 and single guide RNA (sgRNA) complex increases the potential of therapeutics using genome engineering and this can be more efficient if we avoid or reduce the possibility of any unwanted off- target mutations at sites with sequence homology to the on-target, i.e. gene targeted with cas9 system. Therefore, the mechanism by which Cas9 locates specific 20-base-pair (bp) target sequence within a large genome and induces sequence-specific double-stranded DNA (dsDNA) cleavage, is a critical question not only in CRISPR biology but also towards the effort to build a more precise and efficient Cas9-genome editing tool that can be used for therapeutic purpose. Our understanding of the CRISPR-Cas9 process has greatly benefited from extensive structural and biochemical research on Cas9 at various DNA target types, protospacer adjacent motif (PAM) variants, and Cas9 orthologs with their compatible single guide RNA. This review briefly discusses the biology of CRISPR-Cas9 technology before focusing on the most recent structural aspects and providing mechanistic insights into the molecular mechanisms of RNA-guided DNA targeting, cleavage, and repair by the most well-studied Cas9 in genome engineering, the Streptococcus pyogenes (SpCas9 enzyme). By using different biophysical tools it is possible to know the actual binding parameters and kinetics in each step of its action.

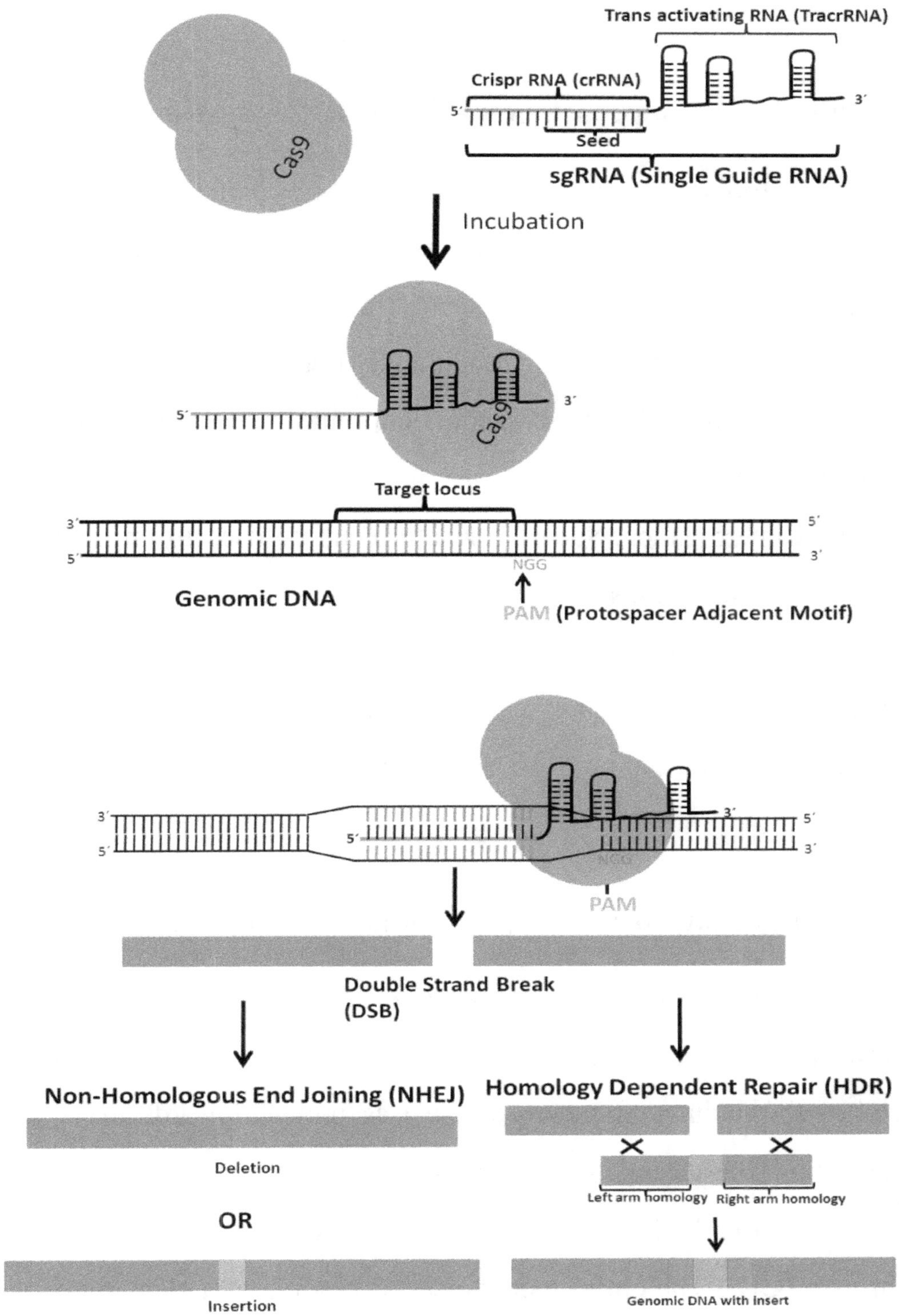

Figure 1.5. Cas9 protein and single guide RNA to form Cas9.sgRNA effector complex (b)Interaction of genomic target DNA and effector complex causing double stranded break followed by repair mechanism pathway like NHEJ and HDR.

1.3.1. Biophysical approach in the CRISPR-Cas system

Biophysical tools provide a means to study the mechanism of the cas9 system in detail. By probing the mechanism in depth one can figure solutions regarding certain limitations of the CRISPR Cas system. There are some limitations of CRISPR-Cas9 for using it as a genome editing tool, mainly being the off-target activity i.e. cleavage at nonspecific sites. Biophysical tools combined with biochemical studies can help gain insight into the cas9 insight such that it can help identify rate limiting steps in the Cas9 DNA interrogation. Such information can be used to improve the specificity of the system. Apart from specificity spcas9 is limited for a single type of PAM (NGG). One can look for other orthologous of cas9 that have different PAM sites and used biophysical techniques to analyse their binding and kinetic parameters that can provide valuable information about their activity and specificity.

1.3.2. Biophysical tools for understanding CRISPR-Cas9 mechanism

Few biophysical tools have been used to analyze the Crispr-Cas9 mechanism. The review briefly discusses the various biophysical tools which people have reported to decipher the binding efficiency of the Cas protein towards the specific targets and nonspecific off-targets like-Forster Resonance Energy Transfer (FRET), beacon assay, single molecule FRET (smFRET) and Bio-Layer Interferometry (BLI), Circular dichroism (CD). The entire mechanism of binding and cleaving of the target region using CRISPR Cas9 can be addressed via these biophysical applications.

Forster resonance energy transfer (FRET) is a quantum process that occurs between two dyes and involves the transmission of excitation from a donor to an acceptor fluorophore through the interaction of two dipoles.

Two fluorophores, donor and acceptor staying at close proximity (8 to 10 nanometers or less) improves spatial resolution of FRET. When FRET occurs, the intensity of the donor fluorescence, lifetime, and quantum efficiency decreases but the fluorescence intensity of the acceptor molecule increases. FLIM-FRET, acceptor photobleaching and sensitized emission are the three different methods by which FRET can be measured. First, the emission spectrum of the donor must overlap the excitation spectrum of the acceptor, and second, the donor and acceptor must be oriented properly for the phenomenon to occur. Bio-Layer Interferometry (BLI), a different tool, is an optical analyzer method that finds the interference pattern of white light. An increase in optical thickness occurs at the biosensor tip as a result of the binding of an immobilised ligand to the surface of the biosensor tip and an analyte in solution. A shift in the interference pattern that may be observed in real time results from a change in the amount of molecules attached to the biosensor tip. Interactions are recorded in real-time, allowing for the monitoring of binding specificity, association rates, and dissociation rates. A dye that fluoresces in the presence of a complementary target is connected to the molecular beacon's 5' end, and a dye that acts as a quencher is covalently attached to the beacon's 3' end. When the beacon is in closed loop, the quencher inhibits the fluorophore from producing light (Fig 1.7). These are single-stranded, hairpin-shaped oligonucleotide probes that have a complementary 18–30 base pair region to the target sequence. Such interactions can also be researched using molecular beacons.

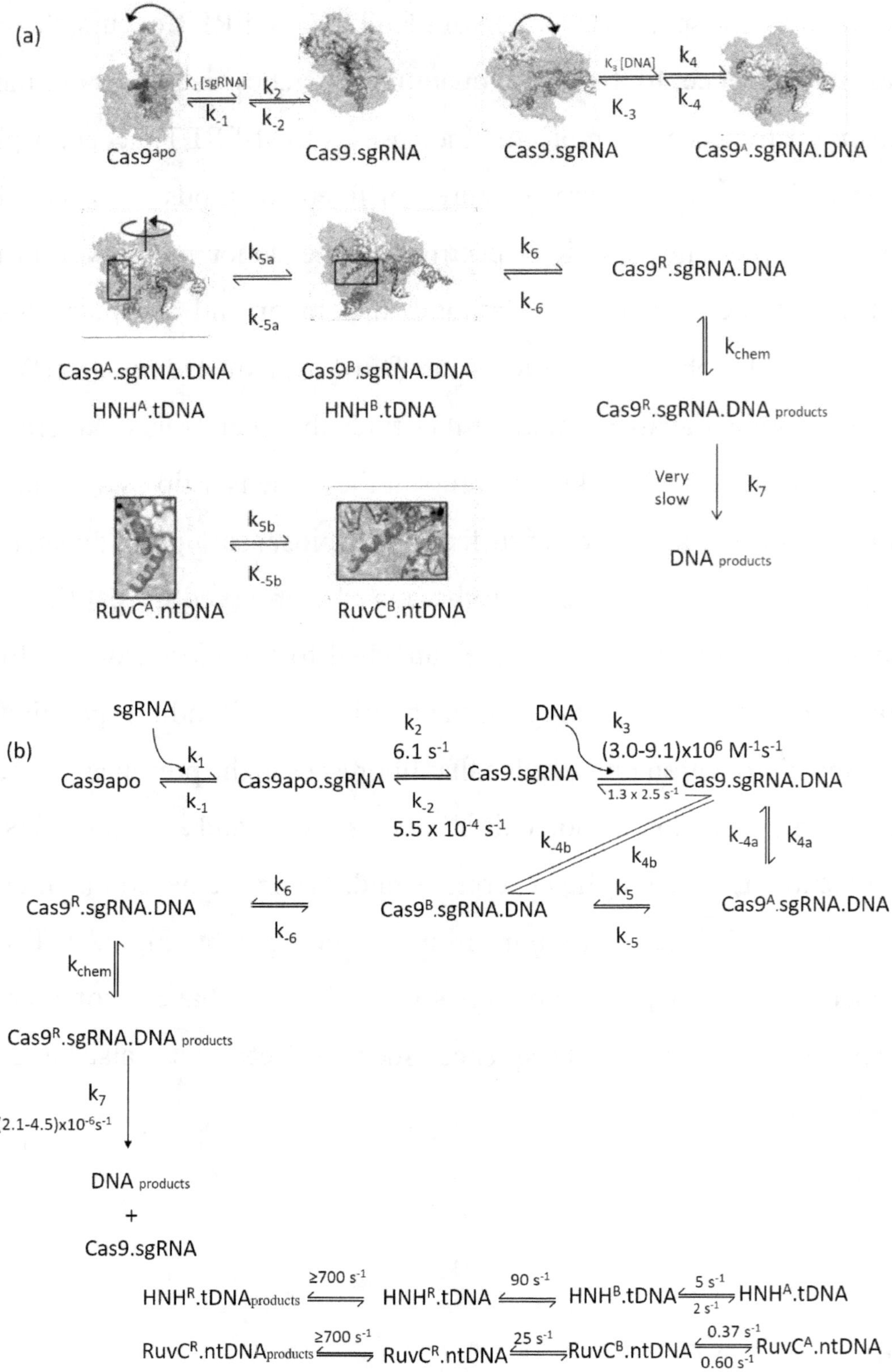

(a)
K1 [sgRNA]
k-1
k2
k-2
Cas9apo
Cas9.sgRNA
K3 [DNA]
K-3
k4
k-4
Cas9.sgRNA
Cas9A.sgRNA.DNA
k5a
k-5a
k6
k-6
Cas9R.sgRNA.DNA
kchem
Cas9A.sgRNA.DNA
HNHA.tDNA
Cas9B.sgRNA.DNA
HNHB.tDNA
Cas9R.sgRNA.DNA products
Very slow
k7
DNA products
k5b
K-5b
RuvCA.ntDNA
RuvCB.ntDNA
(b)
sgRNA
k1
k-1
Cas9apo
Cas9apo.sgRNA
k2
6.1 s-1
k-2
5.5 x 10-4 s-1
Cas9.sgRNA
DNA
k3
(3.0-9.1)x106 M-1s-1
1.3 x 2.5 s-1
Cas9.sgRNA.DNA
k-4b
k4b
k-4a
k4a
Cas9R.sgRNA.DNA
k6
k-6
Cas9B.sgRNA.DNA
k5
k-5
Cas9A.sgRNA.DNA
kchem
Cas9R.sgRNA.DNA products
k7
(2.1-4.5)x10-6s-1
DNA products
+
Cas9.sgRNA
HNHR.tDNAproducts ≥700 s-1 HNHR.tDNA 90 s-1 HNHB.tDNA 5 s-1 2 s-1 HNHA.tDNA
RuvCR.ntDNAproducts ≥700 s-1 RuvCR.ntDNA 25 s-1 RuvCB.ntDNA 0.37 s-1 0.60 s-1 RuvCA.ntDNA

Figure 1.6 (a) Structural basis for conformational changes and kinetic event observed forCas9 binding to sgRNA (Red colored). Green coloured REC lobe restructures upon binding of sgRNA with Cas9. Conformational changes in structure and kinetics observed for Cas9·sgRNA binding to DNA (tDNA and ntDNA coloured in black and cyan). The HNH domain is shown as yellow and RuvC domain is purple. (Cas9^{A}·sgRNA·DNA) is an earlystage and (Cas9^{B}·sgRNA·DNA) is a late stage Cas9 ternary complexes. After the Cas9.sgRNA complex binding to target DNA HNH (yellow) domain swings inward with afast kinetics and then it rotates 180° and moves to its active site which is proximal to the tDNA (black) and the two helices in the RuvC domain (purple) restructures into a single extended helix. (b) Proposed kinetic mechanism for Cas9.sgRNA effector complex formation, binding to target DNA and DNA cleavage with the data generated by FRET experiment.

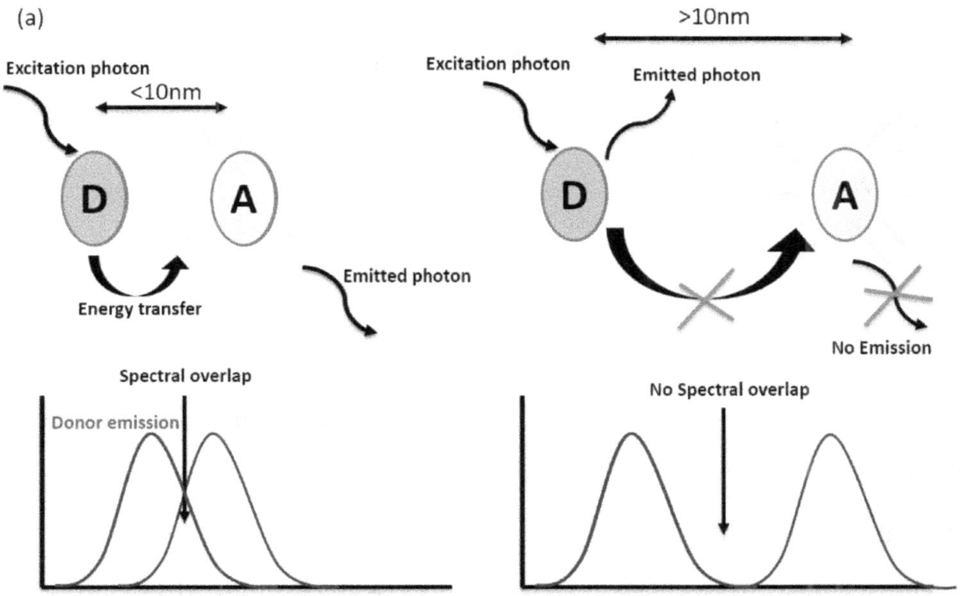

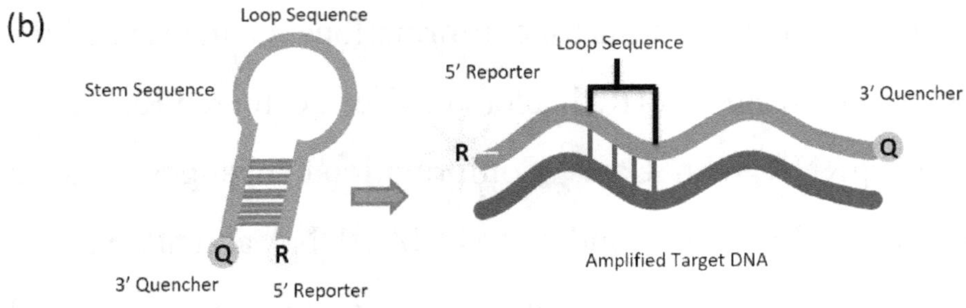

Figure 1.7. (a) Basic principle of FRET study. In the first case, when the donor and acceptor in a close proximity FRET occurs and fluorescence intensity of acceptor increases. But inthe second case when the donor and acceptor are more distant, FRET cannot occur and dueto excitation of donor only donor will emit at some different wavelength whereas there willbe no emission from the acceptor. (b) Concept of molecular beacon. When the beacon is ina loop there will be no fluorescence as quencher will quench the fluorescence of reporter dye. But, when the beacon probe is in open chain bound with some target DNA then the reporter will fluoresce as no quenching will happen there.

1.3.3. Perspective

The molecular mechanism and binding affinity of the conformational changes of the Cas9 protein occurring during its binding with sgRNA, formation of the CRISPR Cas9 effector complex followed by PAM motif recognition by the complex, and complex-mediated target DNA binding and cleavage have been clarified by various biophysical and structural studies. This clarifies how Cas9 can have both the high efficiency and specificity that makes it such a powerful genome editing tool. Despite these advancements in knowledge, it is still unclear what controls the precision and accuracy with which Cas9 recognises DNA targets and what mechanisms guard against unwanted off-target cleavage. It is known that the limiting factor of CRISPR-Cas9 to succeed therapeutics field is it's off target cleavage potency where it can tolerate even three to five base pair mismatches in the PAM distal part of. This problem can be solved by modulating Cas9 conformational dynamics through protein-engineering that can regulate the cleavage domains of cas9 i.e. the HNH and RuvC such that these are active only at the specific sites and not at the off target sites. Biophysical and structural studies can then be used to confirmthe engineered protein conformational dynamics and cleavage at both the specific and off target DNA regions. Our comprehension of the precise processes behind the targeting and cleavage actions of the CRISPR-Cas9 system will continue to be aided by the use of single molecule FRET studies of tagged molecule on target DNA unwinding. The most crucial factors for the CRISPR Cas9 machinery to recognise its eukaryotic genomic target are improved guide RNA design and more precise off target prediction of Cas9 binding or cleavage location. Additionally, we anticipate that the Cas9-mediated genome engineering will significantly increase when more Cas9 orthologs are found and studied. The best candidate of Cas family can be used for genome editing purpose which has highest specificity to target DNA along with efficient binding and cleaving abilities. Utilizing the different proteins of the Cas family we can build a new pipeline andalgorithm for the sequential use of data generated from various biophysical tools to monitor the different kinetics and binding affinity parameters using different DNA targets and off targets. Recent structural, binding, and kinetic investigations offer not only a fundamental comprehension of CRISPR-Cas9 mechanisms but also a framework for structure-based design intended to increase CRISPR-Cas9 effectiveness and reduce off-target effects for human health and treatments.

Chapter 2

Development of a biophysical assay to investigate the quantitative measurement of binding affinity of Cas9 protein and its orthologues

2.1. Introduction:

Microscale Thermophoresis (MST) was carried out as previously described (Wienken et al. 2010). As previously stated, the dSpCas9-GFP protein was purified. The sgRNAs for both spCas9 was IVT synthesised in bulk and then resolved on 8% native PAGE. Then thesgRNAs were PAGE-purified by crush and soaking methods (Petrov, Tsa, and Puglisi 2013). The catalytically inactive/dead Cas9 (dCas9) protein and sgRNA were reconstitutedas equimolar in the final buffer by incubating at 25°C for 10 minutes to make (dCas9-guideRNA) RNP complex. In the final buffer, the HPLC purified 30 bp dsDNA target was incubated with 180 nM RNP complex at 37°C for 60 minutes with multiple concentrations(ranging from 0.09nM to 30M). Monolith.NT (NanoTemper Technologies GmbH, Germany, Munich) was used to determine the RNP complexes' binding affinities to a dsDNA target. The sample was placed into standard treated capillaries from NanoTemper,and measurements were taken at 25°C in blue filter (excitation wavelength 465-490nm; emission wavelength 500-550nm) and with 40% LED power and 40% MST power. For each measurement, three technical replicates were carried out. The OriginPro 8.5 softwareand NanoTemper analytic tools were used to analyse the data.

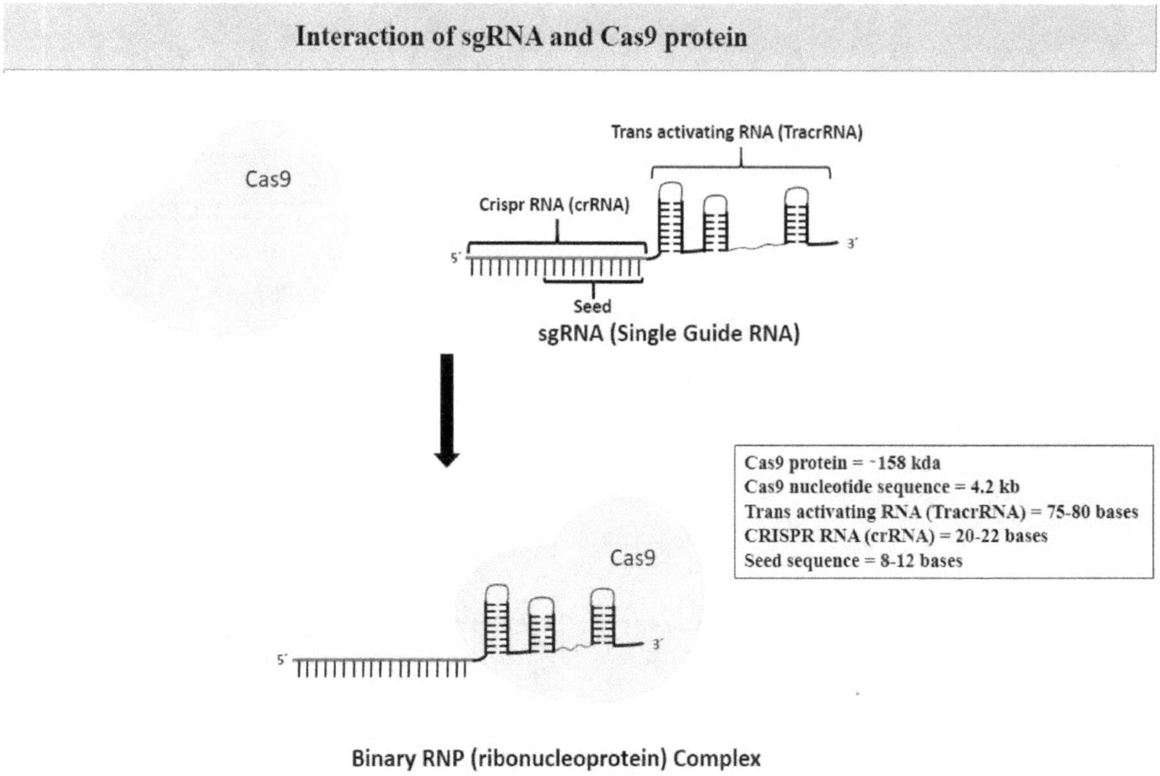

Figure 2.1: Binary complex was made using Cas9 protein and sgRNA.

Figure 2.2: Ternary complex was made using RNP complex and DNA target.

Binding of RNP complex and DNA Target

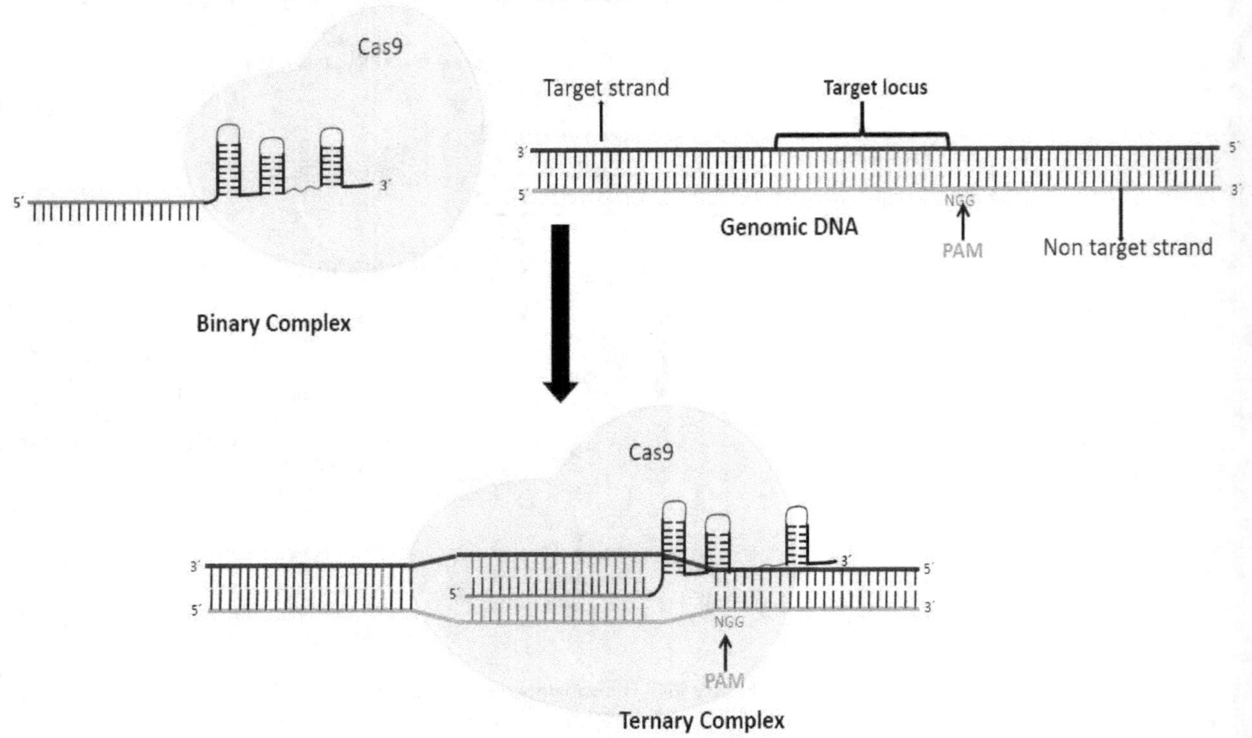

Figure 2.2: Ternary complex was made using RNP complex and DNA target.

2.2. Materials and Methods

2.2.1. Construction of plasmids for SpCas9, FnCas9, CjCas9, SaCas9, dSpCas9, dFnCas9, dCjCas9, GFP fused dSpCas9, dCjCas9, dSaCas9, dFnCas9:

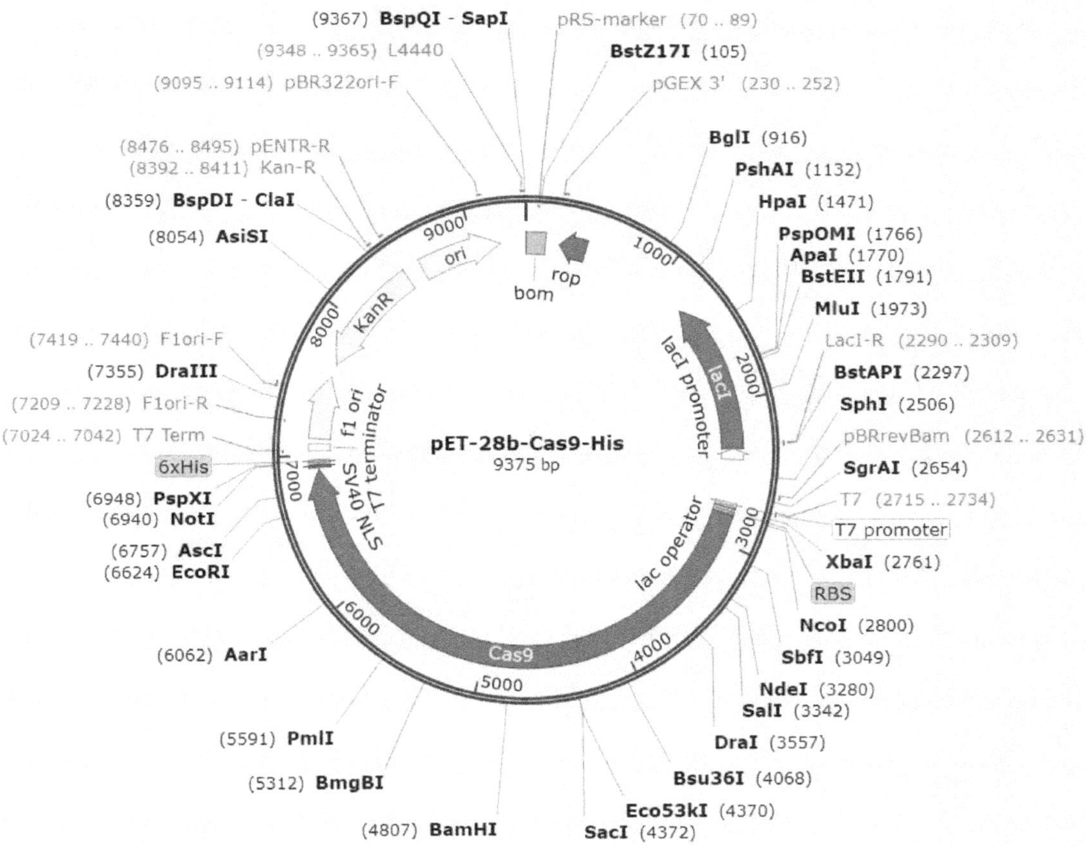

WTSpCas9

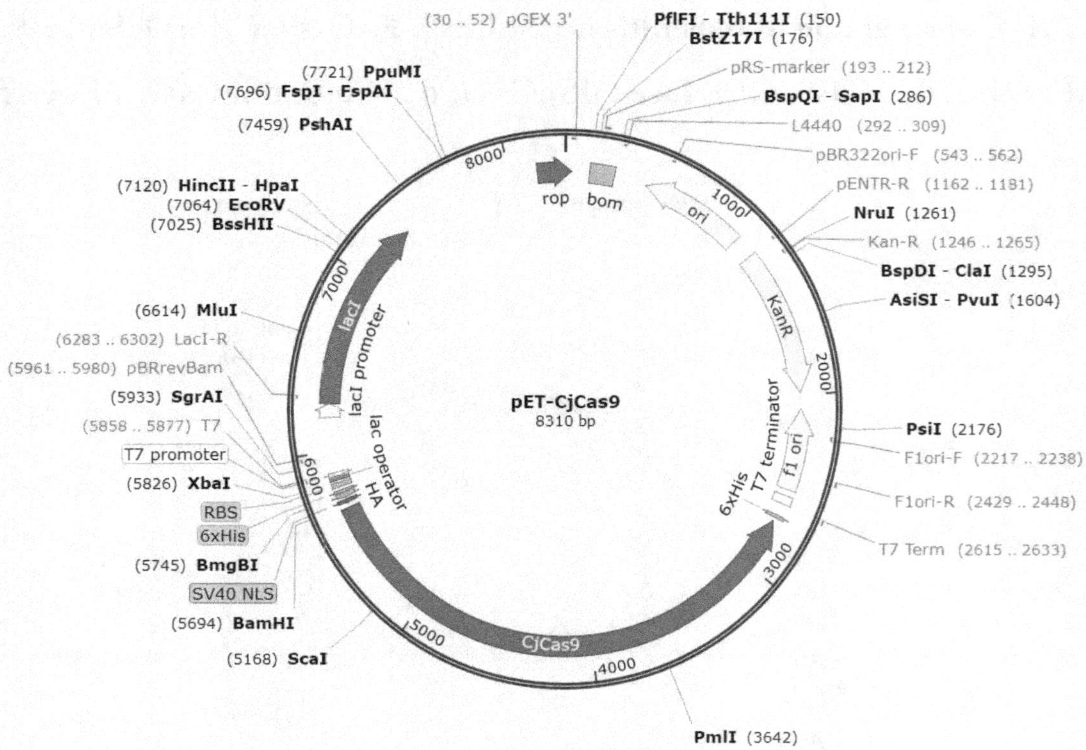
(30 .. 52) pGEX 3'
PflFI - Tth111I (150)
BstZ17I (176)
pRS-marker (193 .. 212)
BspQI - SapI (286)
L4440 (292 .. 309)
pBR322ori-F (543 .. 562)
pENTR-R (1162 .. 1181)
NruI (1261)
Kan-R (1246 .. 1265)
BspDI - ClaI (1295)
AsiSI - PvuI (1604)
PsiI (2176)
F1ori-F (2217 .. 2238)
F1ori-R (2429 .. 2448)
T7 Term (2615 .. 2633)
PmlI (3642)
(5168) ScaI
(5694) BamHI
SV40 NLS
(5745) BmgBI
6xHis
RBS
(5826) XbaI
T7 promoter
(5858 .. 5877) T7
(5933) SgrAI
(5961 .. 5980) pBRrevBam
(6283 .. 6302) LacI-R
(6614) MluI
(7025) BssHII
(7064) EcoRV
(7120) HincII - HpaI
(7459) PshAI
(7696) FspI - FspAI
(7721) PpuMI
rop
bom
ori
KanR
f1 ori
T7 terminator
6xHis
CjCas9
HA
lac operator
lacI promoter
lacI
pET-CjCas9
8310 bp
1000
2000
3000
4000
5000
6000
7000
8000

WTCjCas9

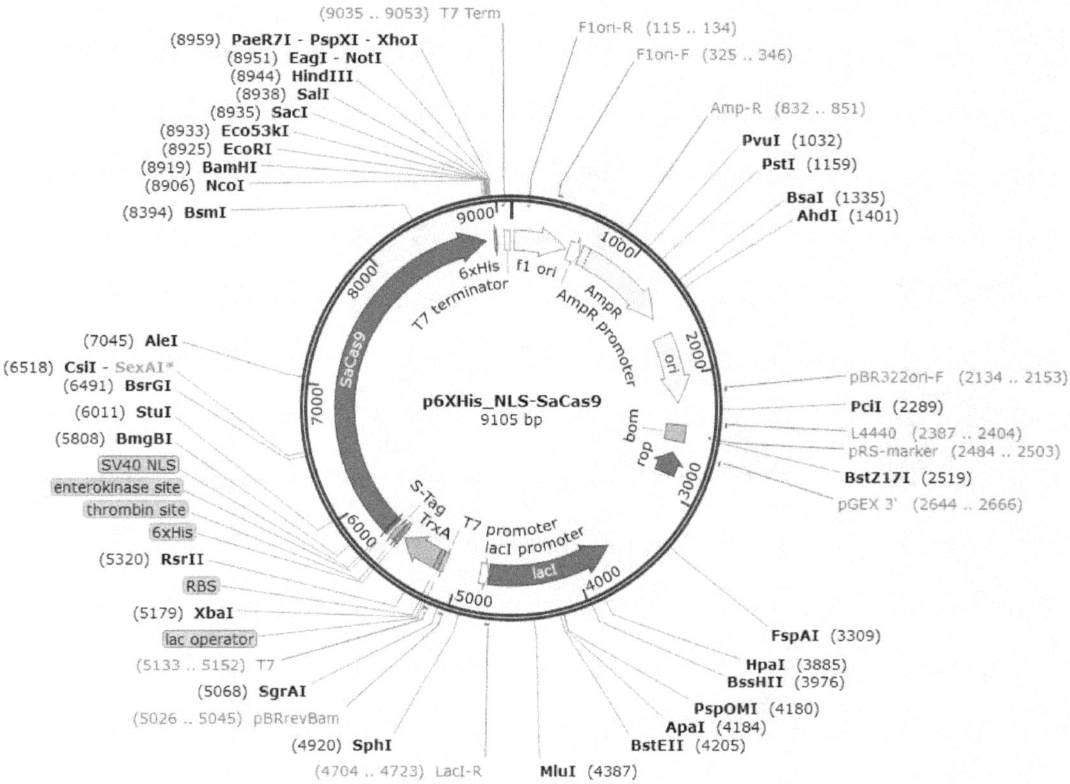
(9035 .. 9053) T7 Term
(8959) PaeR7I - PspXI - XhoI
(8951) EagI - NotI
(8944) HindIII
(8938) SalI
(8935) SacI
(8933) Eco53kI
(8925) EcoRI
(8919) BamHI
(8906) NcoI
(8394) BsmI
F1ori-R (115 .. 134)
F1ori-F (325 .. 346)
Amp-R (832 .. 851)
PvuI (1032)
PstI (1159)
BsaI (1335)
AhdI (1401)
pBR322ori-F (2134 .. 2153)
PciI (2289)
L4440 (2387 .. 2404)
pRS-marker (2484 .. 2503)
BstZ17I (2519)
pGEX 3' (2644 .. 2666)
FspAI (3309)
HpaI (3885)
BssHII (3976)
PspOMI (4180)
ApaI (4184)
BstEII (4205)
MluI (4387)
(4704 .. 4723) LacI-R
(4920) SphI
(5026 .. 5045) pBRrevBam
(5068) SgrAI
(5133 .. 5152) T7
lac operator
(5179) XbaI
RBS
(5320) RsrII
6xHis
thrombin site
enterokinase site
SV40 NLS
(5808) BmgBI
(6011) StuI
(6491) BsrGI
(6518) CsiI - SexAI*
(7045) AleI
p6XHis_NLS-SaCas9
9105 bp
6xHis
f1 ori
T7 terminator
AmpR
AmpR promoter
ori
bom
rop
lacI
lacI promoter
T7 promoter
S-Tag
TrxA
SaCas9
1000
2000
3000
4000
5000
6000
7000
8000
9000

WTSaCas9

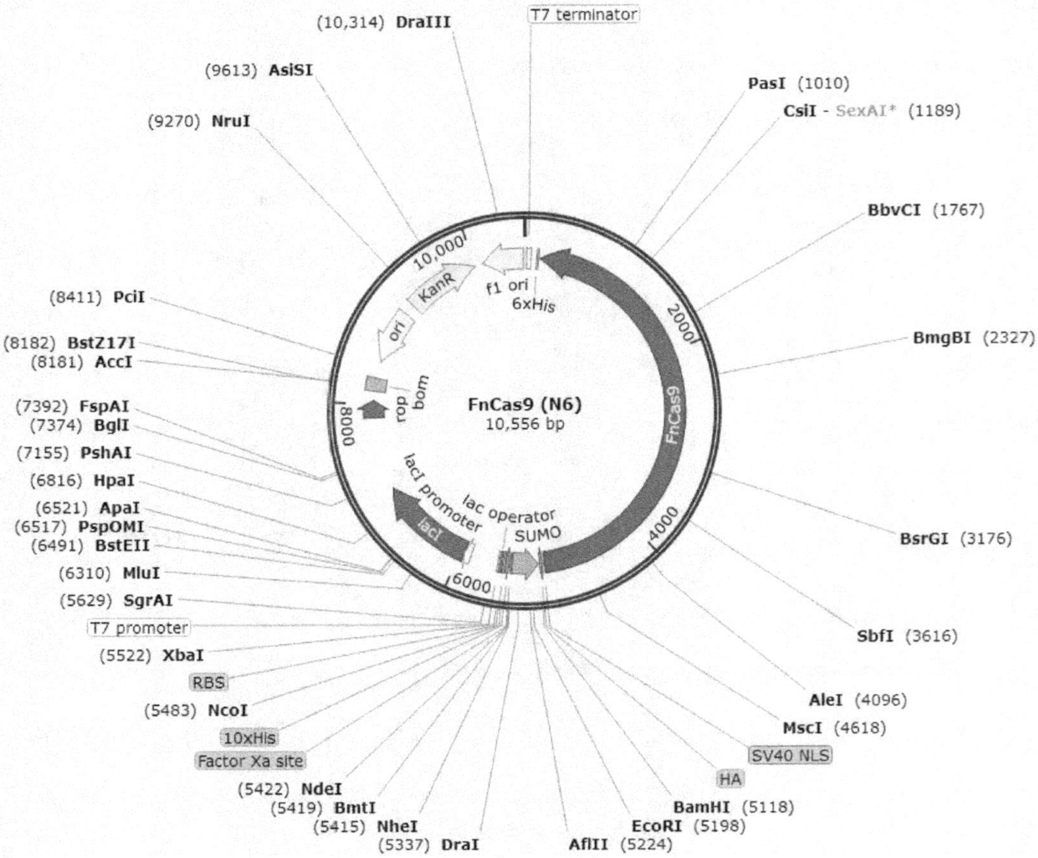

(10,314) DraIII
T7 terminator
(9613) AsiSI
PasI (1010)
CsiI - SexAI* (1189)
(9270) NruI
BbvCI (1767)
10,000
f1 ori
6xHis
KanR
ori
2000
(8411) PciI
BmgBI (2327)
(8182) BstZ17I
(8181) AccI
bom
rop
FnCas9 (N6)
10,556 bp
FnCas9
8000
(7392) FspAI
(7374) BglI
(7155) PshAI
(6816) HpaI
lacI promoter
lac operator
SUMO
lacI
4000
(6521) ApaI
(6517) PspOMI
(6491) BstEII
BsrGI (3176)
(6310) MluI
6000
(5629) SgrAI
T7 promoter
SbfI (3616)
(5522) XbaI
RBS
AleI (4096)
(5483) NcoI
MscI (4618)
10xHis
SV40 NLS
Factor Xa site
HA
(5422) NdeI
BamHI (5118)
(5419) BmtI
EcoRI (5198)
(5415) NheI
(5337) DraI
AflII (5224)

WTFnCas9(N6)

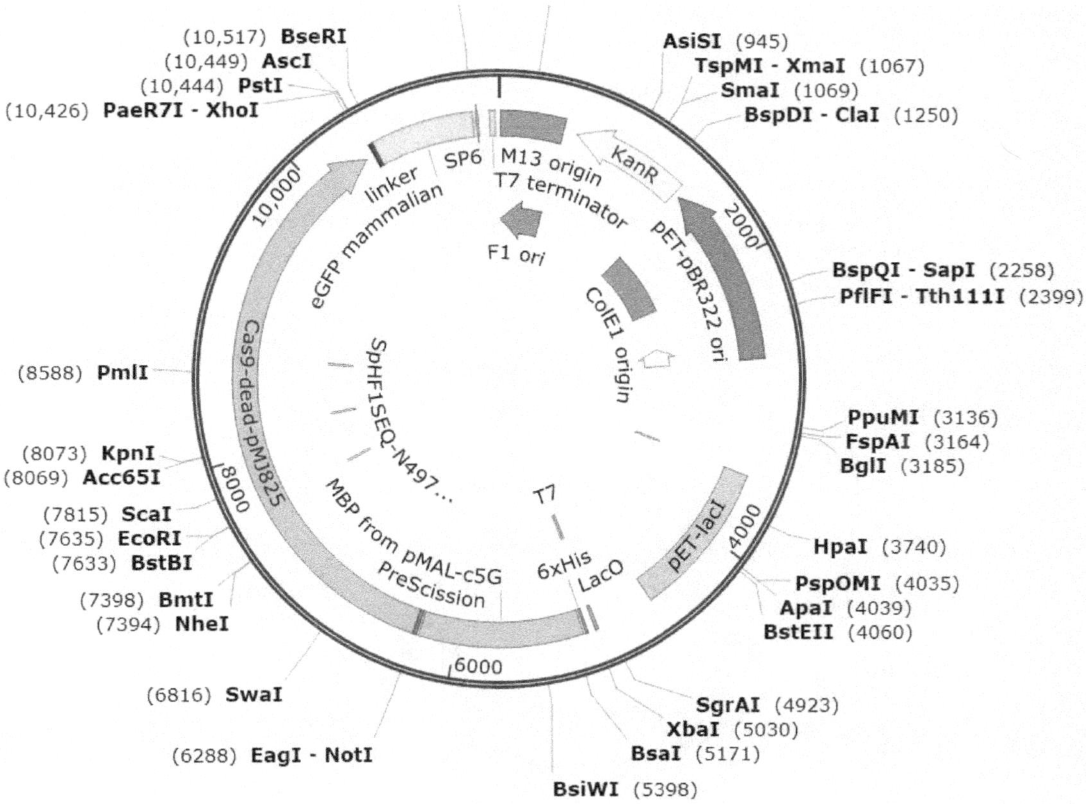
(10,517) BseRI
(10,449) AscI
(10,444) PstI
(10,426) PaeR7I - XhoI
AsiSI (945)
TspMI - XmaI (1067)
SmaI (1069)
BspDI - ClaI (1250)
BspQI - SapI (2258)
PflFI - Tth111I (2399)
PpuMI (3136)
FspAI (3164)
BglI (3185)
HpaI (3740)
PspOMI (4035)
ApaI (4039)
BstEII (4060)
SgrAI (4923)
XbaI (5030)
BsaI (5171)
BsiWI (5398)
(6288) EagI - NotI
(6816) SwaI
(7398) BmtI
(7394) NheI
(7815) ScaI
(7635) EcoRI
(7633) BstBI
(8073) KpnI
(8069) Acc65I
(8588) PmlI
SP6
M13 origin
T7 terminator
KanR
F1 ori
pET-pBR322 ori
ColE1 origin
linker
eGFP mammalian
Cas9-dead-pMJ825
SpHF1SEQ-N497...
MBP from pMAL-c5G
PreScission
6xHis
LacO
T7
pET-lacI
2000
4000
6000
8000
10,000

dSpeGFP

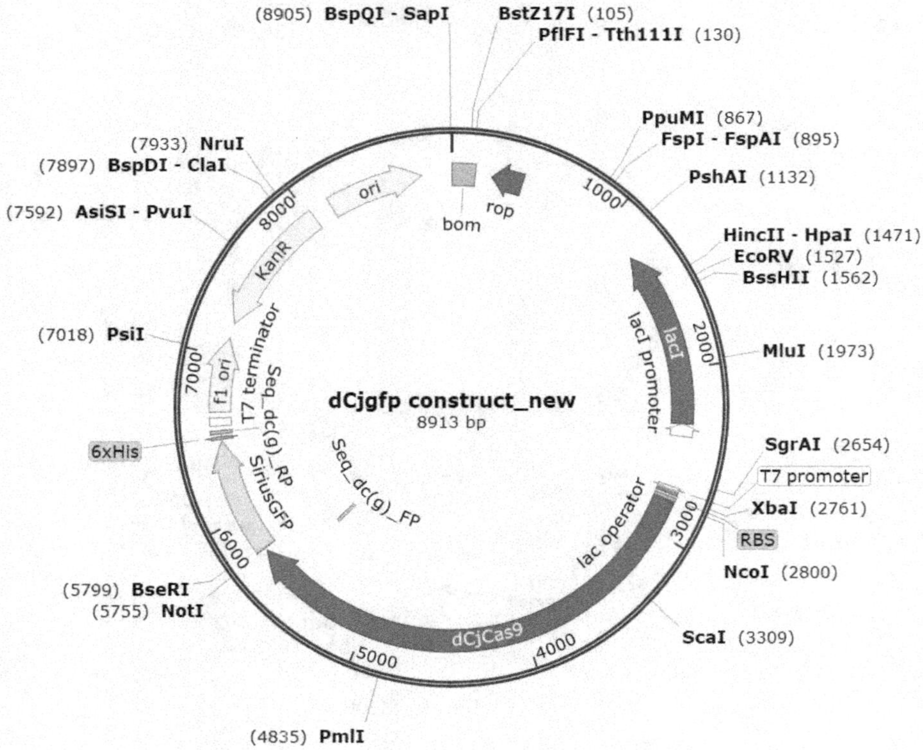

(8905) BspQI - SapI
BstZ17I (105)
PflFI - Tth111I (130)
PpuMI (867)
FspI - FspAI (895)
PshAI (1132)
HincII - HpaI (1471)
EcoRV (1527)
BssHII (1562)
MluI (1973)
SgrAI (2654)
T7 promoter
XbaI (2761)
RBS
NcoI (2800)
ScaI (3309)
(4835) PmlI
(5755) NotI
(5799) BseRI
6xHis
(7018) PsiI
(7592) AsiSI - PvuI
(7897) BspDI - ClaI
(7933) NruI
ori
bom
rop
lacI
lacI promoter
lac operator
dCjCas9
SiriusGFP
Seq_dc(g)_RP
Seq_dc(g)_FP
T7 terminator
f1 ori
KanR
dCjgfp construct_new
8913 bp
1000
2000
3000
4000
5000
6000
7000
8000

dCjGFP

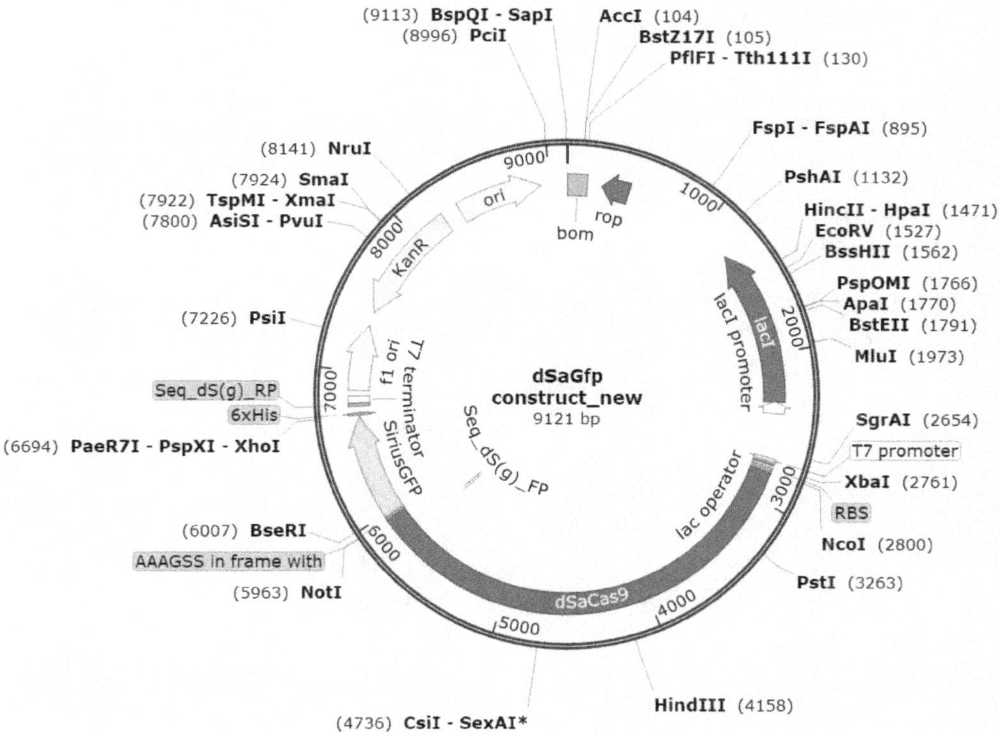
(9113) BspQI - SapI
(8996) PciI
AccI (104)
BstZ17I (105)
PflFI - Tth111I (130)
FspI - FspAI (895)
PshAI (1132)
HincII - HpaI (1471)
EcoRV (1527)
BssHII (1562)
PspOMI (1766)
ApaI (1770)
BstEII (1791)
MluI (1973)
SgrAI (2654)
T7 promoter
XbaI (2761)
RBS
NcoI (2800)
PstI (3263)
HindIII (4158)
(4736) CsiI - SexAI*
(5963) NotI
AAAGSS in frame with
(6007) BseRI
(6694) PaeR7I - PspXI - XhoI
6xHis
Seq_dS(g)_RP
(7226) PsiI
(7800) AsiSI - PvuI
(7922) TspMI - XmaI
(7924) SmaI
(8141) NruI
ori
bom
rop
KanR
lacI
lacI promoter
lac operator
dSaCas9
SiriusGFP
T7 terminator
f1 ori
Seq_dS(g)_FP
dSaGfp
construct_new
9121 bp
1000
2000
3000
4000
5000
6000
7000
8000
9000

dSaGFP

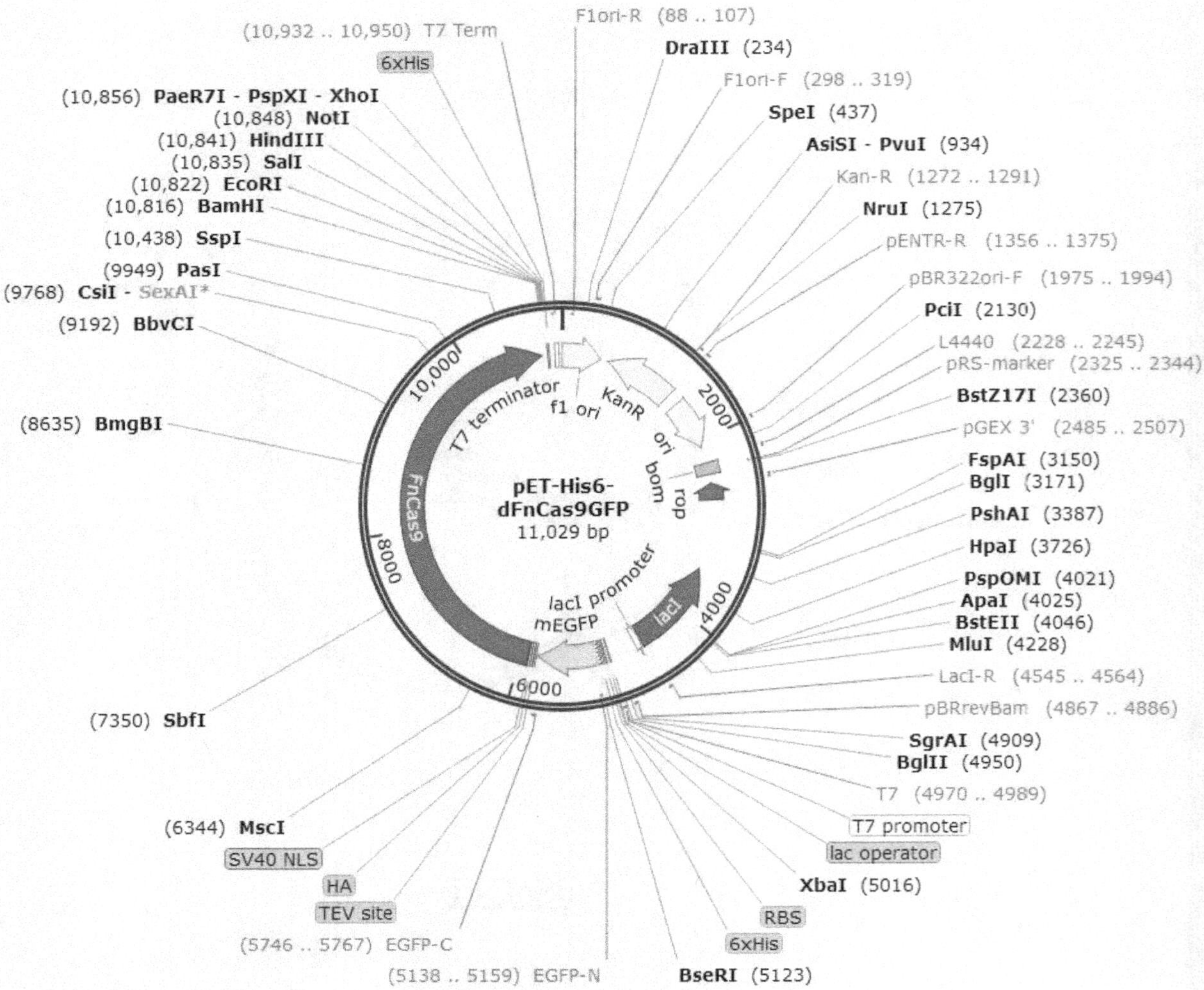

dFnGFP

2.2.2. Purification of different Cas9 proteins

Protein purification (SpCas9, dSpCas9-GFP, dSpCas9, FnCas9, dFnCas9, and dFnCas9- GFP, SaCas9, dSaCas9-GFP, SaCas9, CjCas9, dCjCas9, dCjCas9-GFP) was carried out as described followingly. In a nutshell, plasmids encoding several Cas9 proteins were transformed and expressed in E. coli Rosetta2 (DE3) from Novagen. Rosetta2 (DE3) cellswhich are the protein expressing cells were cultured in 10 ml of Luria Bertani (LB) media(supplemented with respective antibiotic) as primary culture and incubated at 37°C with the rotation (180-200rpm) for overnight. Post incubation secondary cultures were inoculated for each plasmid with 1% of primary culture and incubated at 37°C with therotation (180-200 rpm) until OD600 (around 1-2 hour) reached in between 0.4-0.6.

Protein expression was induced by the addition of isopropyl β-D-thiogalactopyranoside (IPTG) to the secondary culture and further incubated overnight at 18°C with the rotation (180-200 rpm). IPTG concentration varied as per the plasmid inoculated for expressing protein. Next day, the bacterial cells were harvested by centrifugation at 5000 rpm for 10 min at 4°C. The supernatants were discarded and the pellets were washed with 1X PBS. Washed pellets were resuspended in freshly prepared lysis buffer (20mM HEPES, pH- 7.5, 500mM NaCl, 5% glycerol, 1mM DTT, 100 μg /ml lysozyme) supplemented with 1X Protease inhibitor cocktail (PIC, Roche) and 2mM phenylmethylsulfonyl fluoride (PMSF). Along with chemical lysis, the mechanical lysis was done through a sonication process, with conditions of 5 second on and 30 second off with amplitude of 35%, for 5- 10 mins or until the solution turned black in colour from milky white. The sonicated solutions were transferred into the oakridge tube and centrifuged at 18-20k rpm for 1 hourto get cell debris in the pellet. Affinity cOmplete his-tag purification Nickel-NTA beads (Roche) were taken in required quantity into the column and washed with 3 column volumes of double distilled water followed by 2 column volumes of washing buffer (20mM HEPES pH-7.5, 500mM NaCl and 5mM imidazole). Supernatant obtained from centrifugation of the cell lysate was added to these prepared beads and incubated with gentle rotation in a cold room (4°C) for 1-2 hours. After this incubation, the beads and protein solution slurry were poured into a column where the supernatant was discarded with the flow through and the beads remained in the column. The beads with attached proteins were washed with 10 times the column volumes of the wash buffer. Next, the elution buffer (20mM HEPES pH-

7.5, 500mM NaCl and 500mM imidazole) was added to the beads and incubated with gentle rotation in a cold room (4°C) for 45-60 mins. Afterincubation, the elution buffer along with protein was eluted into fresh falcon tubes. In next step, the elution buffer with protein was exchanged and concentrated with final (storage) buffer (20mM HEPES pH-7.5, 150mM KCl, 10mM MgC12, 1mM DTT/ TCEP and 5-10% glycerol) using Amicon centricon filters (Merk, 100-Kda MW cut-Off) at 4000 rpm in 4°C with 3-5 times the column volume. The protein in the final (storage) buffer was resolved on the 8% SDS-PAGE with Coomassie brilliant blue stain to confirm and check the integrity. The SUMO tag from FnCas9 was deleted with Ulp1 protease, whereas the 6 His-MBP tag from dSpCas9-EGFP was removed by incubating it overnightin cleavage buffer (50 mM TrisCl, pH 8, 150 mM NaCl, 1 mM EDTA, 1 mM DTT).

Following SDSPAGE confirmation, the protein was purified further using a HiLoad Superdex 200 16/600 column (GE Healthcare) in the final (storage) buffer, and the sample tubes were collected according to the run's chromatogram. The protein in varioustubes was resolved on an 8 percent SDS-PAGE with coomassie brilliant blue dye to check its size and integrity. The best-quality protein samples vials were combined and aliquoted into tiny amounts in microfuge tubes. Purified proteins were quantified using aPierce BCA protein assay kit (Thermo Fisher Scientific) and kept at -80°C until needed.

2.2.3. Preparation of HiLoad Superdex 200 16/600 size exclusion column

The final storage buffer with the autoclaved milliQ water was prepared according to the composition without 10% glycerol (which may block the column). Both, the autoclaved milliQ and final buffer were filtered with the 0.2μM filter assembly and then degassed by incubating on a sonication bath with cap loosen for 1 hour.

Meanwhile the temperature ofthe column chromatography (AKTA purifier) chamber was set and maintained at 4ºC throughout protein purification. The HiLoad Superdex 200 16/600 column was washed and equilibrated with degassed milliQ water and final buffer as per the manual instructions. The proteins were concentrated with an amicon centricon filter in ~6ml withmaximum 2 mg/ml concentration and filtered with 0.2μM syringe filter before loading onto the column. The filtered protein was loaded onto the AKTA purifier as instructed.

The purification of protein was done automatically by setting the program with all the necessary details. The proteins were separated and collected into the collection tubes asper the chromatograms. The eluted protein aliquots were then resolved on 8% SDS-PAGE to check the size and integrity. The intact and desired sized protein tubes were then pooled and mixed with 10% of glycerol. The concentration of the protein was calculated with BCA kit and allotted into small volumes to store at -80ºC until further use.

2.2.4. Guide RNA design and preparation:

The Guide RNAs for target sequences were designed using the freely available CRISPOR software (http://crispor.tefor.net; Concordet and Haeussler 2018) according to the specifications given online. The hg38 was kept as the reference human genome in the background to check for any possible off targets of designed guide RNA. The software is yet not updated with FnCas9 specific parameters but since the SpCas9 and FnCas9 both require the same PAM sequence (5'-NGG-3') and the length of crRNA, the guides designed for SpCas9 were relevant for both and thus used with FnCas9 as well. For CjCas9 and SaCas9 the PAM sequences were chosen on the basis of CRISPOR software again.

2.2.5. Template preparation for single guide RNA (sgRNA)

In the case of SpCas9, two oligos were designed as forward primer and universal reverseprimer. Forward Primer was 52 nucleotide in length which contains from 5'-3': initial 17nucleotides as minimal T7 RNA polymerase promoter followed by 20 nucleotides complementary to the target DNA site, remaining 15 nucleotides a part of TracrRNA andcomplementarity with the initial 15 nucleotides of the universal reverse primer. The universal reverse primer was 80 nucleotides long, containing the complementary sequence of TracrRNA. A PCR reaction was set up using these oligos to prepare a full- length dsDNA template. Then the dsDNA template was checked on 2% agarose gel and further used in IVT (*In vitro* transcription) reaction or stored at -20°C until further use.

For FnCas9's single guide RNA (sgRNA), the forward primer, as well as the universal reverse primer, were 72 nucleotides long. The forward primer contains from 5'-3': initial17 nucleotides as minimal T7 RNA polymerase promoter followed by 20 nucleotides, complementary to the target DNA site and remaining 35 nucleotides a part of TracrRNAand complementarity with the initial 35 nucleotides of the universal reverse primer. The universal reverse primer was a complementary sequence of TracrRNA. A PCR reaction was set up using these oligos to prepare a full-length dsDNA template. This synthesizedand amplified dsDNA template was checked on 2% agarose gel and further used in IVT reaction or stored at -20°C until further use. For Cj and SaCas9, we took the exactly overlapping primers.

PCR reaction mixture for the synthesis of dsDNA:

Reagents	Volume (μl)
10X PCR buffer	5
dNTPs (2.5mM)	4
Forward primer (10μM)	2.5
Universal reverse primer (10μM)	2.5
Taq polymerase	1
Nuclease free water	Up to 50
Total volume	50

Table 2.2: PCR conditions for SpCas9 sgRNA specific template synthesis:

Temperature (°C)	Time	Cycle
95	3 minutes	1
95	10 sec	30
45	30 sec	
72	30 sec	
72	5 minutes	1

2.1.1. *In vitro* Transcription (IVT) for guide RNA synthesis

Thermo Fisher Scientific's MegaScript T7 Transcription kit was used to *in vitro* generateguide RNAs from a T7 promoter-containing template. IVT reactions were set up according to the directions in the kit handbook and incubated at 37°C overnight. After incubation, DNase treatment at 37°C for 30 minutes was used to remove the dsDNA template from IVT RNA. The reaction was then validated by running RNA on a 2% agarose gel, in which 1 μl of RNA sample was combined with 4 μl of nucleasefree water(NFW) and 5 µl of RNA loading dye, heated at 90-95ºC for 2 minutes, and then snap cooled prior loading onto the gel. The reaction was confirmed by resolving it in a 1 percent agarose gel dyed with EtBr and observed with a Syngene UV transilluminator.

The *in vitro* transcribed guide RNAs were purified using a NucAway spin column (Thermofisher Scientific) following the manufacturer's directions after validation. IVTsgRNAs/crRNAs were kept at -20°C until needed.

***In Vitro* transcription (IVT) reaction mixture:**

Reagents	Volume (µl)
10X Reaction buffer	2
ATP (75mM)	2
CTP (75mM)	2
UTP (75mM)	2
GTP (75mM)	2
Enzyme mix	2
Template	5
Nuclease free water	Up to 20
Total volume	20

Reaction conditions for *in vitro* transcription (IVT):

Addition of 1 µl of Turbo DNase	
37	30 minutes
70	10 minutes

2.1.1. Guide RNA purification

The IVT RNA was by using the NucAway spin column (Thermo Fisher Scientific) as perthe instructions. Briefly, the NucAway spin column was taken and filled with 650 µl of nuclease free water to homogenize the gel bed. The column was vortexed for a few seconds and incubated at room temperature for 5-15 mins. After incubation, the bubbles in the column were removed by tapping and vortexing it; presence of bubbles may entrapproducts resulting in lower yield while recovering. Then the column was placed in the collection tube followed by centrifugation at 750x g for 2 min. Next, the column was transferred into a fresh 1.5ml microfuge tube and the IVT RNA was loaded onto the topcentre of the gel bed without damaging it. The column with IVT product was incubated for 1 min on stand and again centrifuged at 750x g for 2 mins. The RNA product was collected in the final flow through and the column was discarded. The concentration and the purity of RNA was checked with nanodrop (Denovix) and stored at -20°C until further use. For all biophysical assays, sgRNA got purified using 12% Urea Page and was gel checked.

2.2.6. Target DNA Duplex Formation:

DNA duplex was prepared using HPLC purified 30nt FP and RP by annealing them at 95 ^{0}C for 5mins and then cooling down to 25^{0}C at 5^{0}C/min ramp rate. And then it was checkedin the 4% agarose gel.

2.2.7. *In Vitro* Cleavage (IVC) Assay:

The *In vitro* cleavage reactions were performed as follows: Cas9 protein and sgRNA weremixed in equimolar amount (500nM for PCR amplicons while 100nM for linearized plasmids as substrate) in final buffer (20 mM HEPES, pH7.5, 150mM KCl,

1mM DTT, 10% glycerol, 10mM MgCl2) and incubated at 25°C for 10 minutes to make Cas9-guide RNA complex. Then 100ng of DNA substrate (PCR amplicon or linearised plasmid) wasadded to the complex and incubated at 37°C for 30 min-1 hr in total 10-15μl of reaction volume. For plasmid as substrate, it is first linearized with a restriction enzyme (New England Biolabs) as per the provided protocol. The IVC reaction is then treated with 1μl of Proteinase K (20 mg/ml) for 10 minutes at 42°C and heat inactivated at 70°C for 10 minutes. The reaction was checked by resolving it in EtBr-stained 1% agarose gel, visualized by Syngene UV transilluminator and quantified by ImageJ.

2.2.8. Microscale thermophoresis (MST) assay:

MST was carried out as previously described (Wienken et al. 2010). As previously stated,the dSpCas9-GFP protein was purified. The sgRNAs for both spCas9 was IVT synthesised in bulk and then resolved on 8% native PAGE. Then the sgRNAs were PAGE-purified by crush and soaking methods (Petrov, Tsa, and Puglisi 2013). The catalytically inactive/dead Cas9 (dCas9) protein and sgRNA were reconstituted as equimolar in the final buffer by incubating at 25°C for 10 minutes to make (dCas9-guide RNA) RNP complex. In the final buffer, the HPLC purified 30 bp dsDNA target was incubated with 180 nM RNP complex at 37°C for 60 minutes with multiple concentrations (ranging from 0.09nM to 30M). Monolith NT (NanoTemper Technologies GmbH, Germany,Munich) was used to determine the RNP complexes' binding affinities to a dsDNA target. The sample was placed into standard treated capillaries from NanoTemper, and measurements were taken at 25°C in blue filter (excitation wavelength465-490nm; emission wavelength 500-550nm) with 40% LED power and 40% MST power. For each measurement, three technical replicates were carried out. The OriginPro 8.5 software and NanoTemper analytic tools were used to analyse the data.

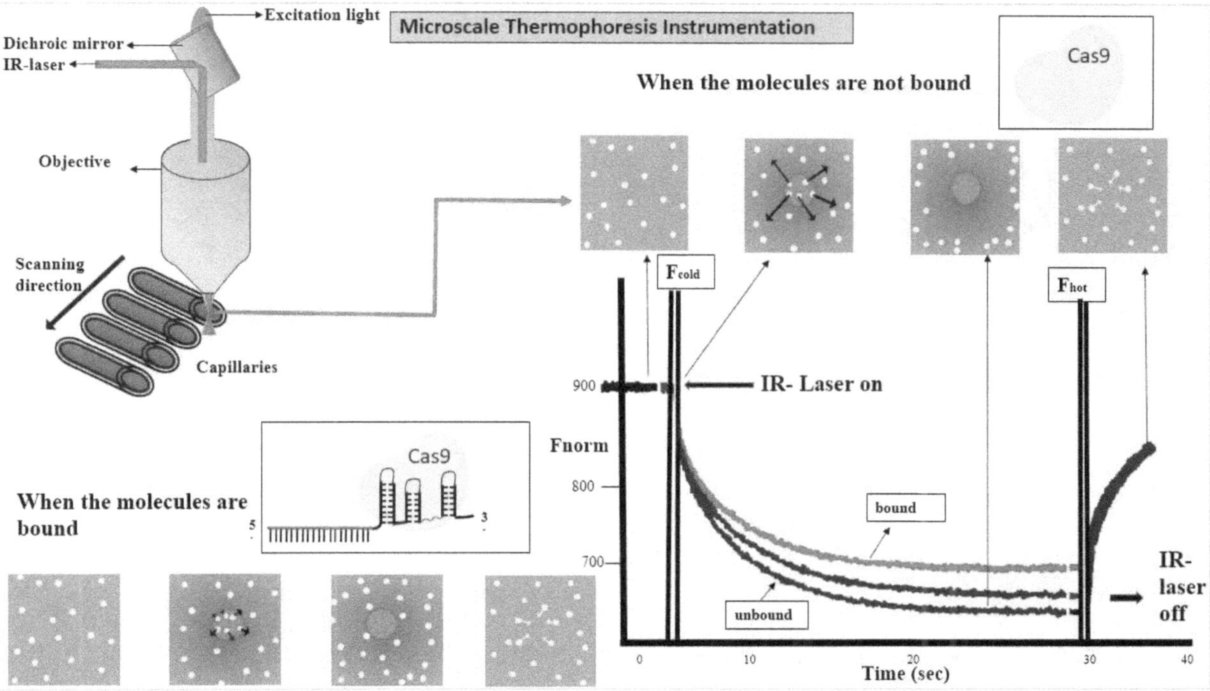

Figure 2.3: Microscale Thermophoresis Instrumentation

2.3. Results and Discussions

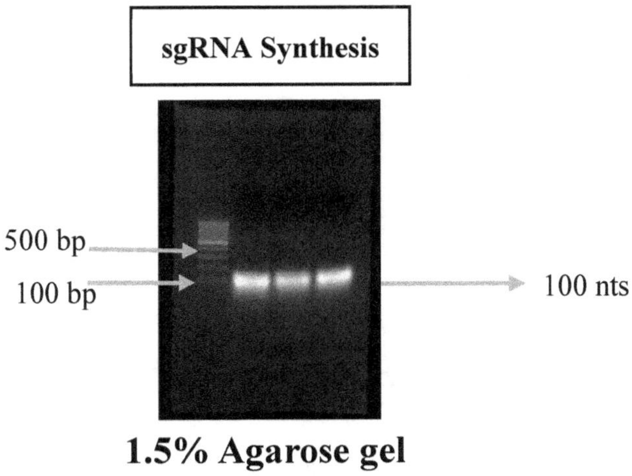

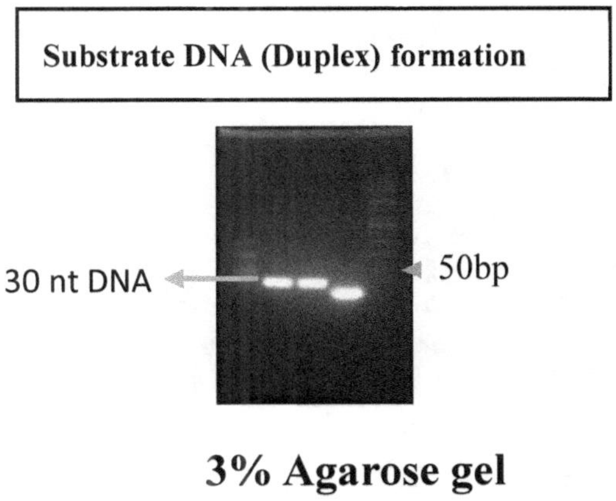

Figure 2.3.4 and 2.3.5: Gel images of sgRNA for SpCas9 and Target DNA duplex

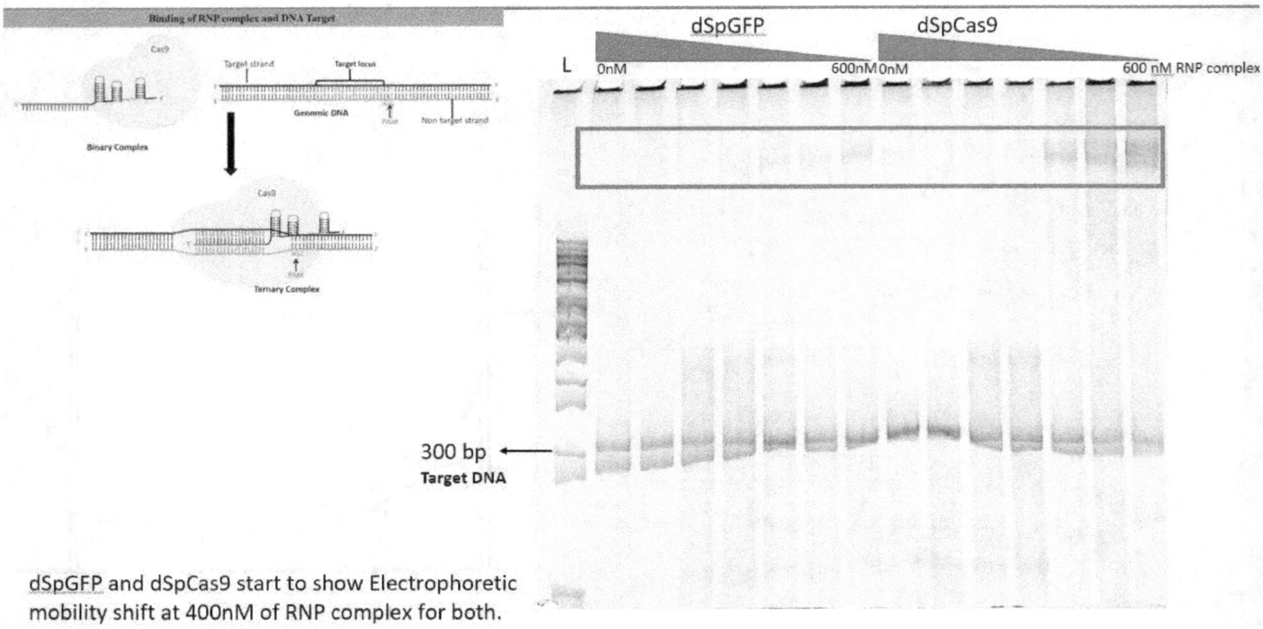

Figure 2.3.6. Experiment for comparing dSpGFP and dSpCas9 made RNP complex withDNA target binding through EMSA.

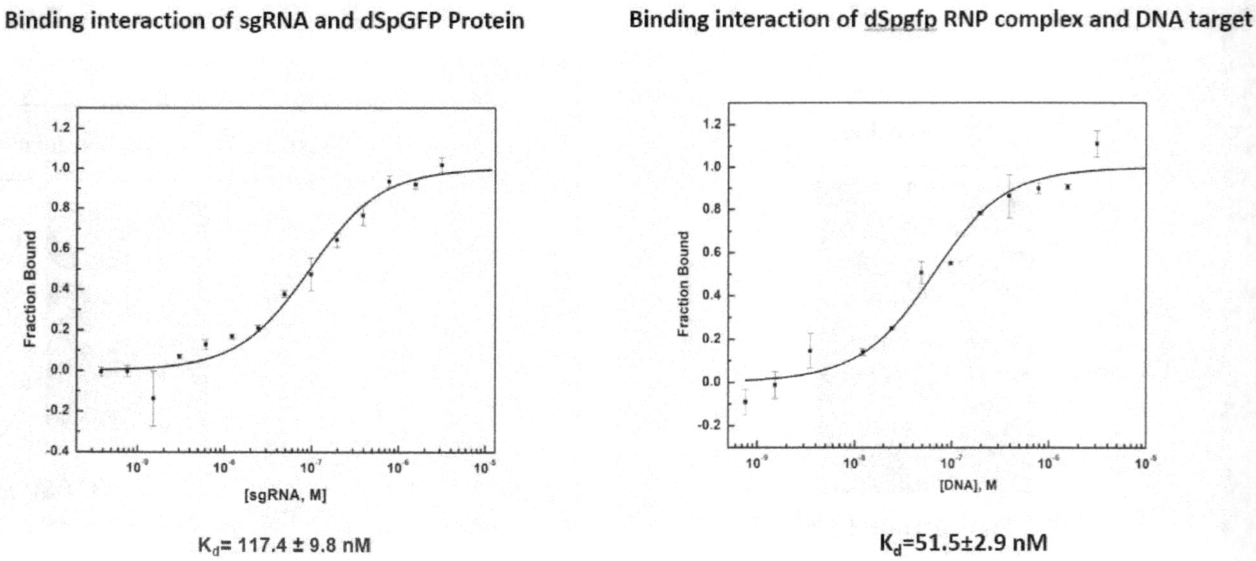

Figure 2.3.7.: Standardization of binding assay for Cas9 system

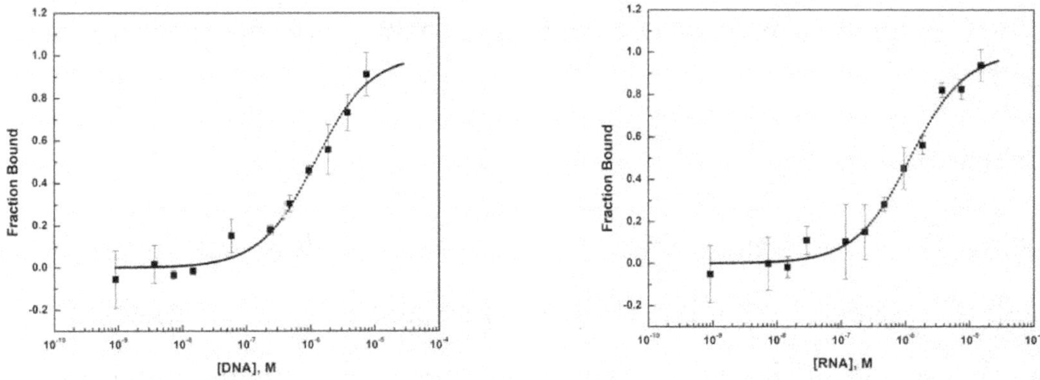

(dSpGFP)Protein against Target DNA

K_d=1423.55+/-36.15nM

(dSpGFP)Protein made RNP complex against non compatible Target DNA

Kd=1637.15+/-19.35 nM

Figure 2.3.8. Control Experiment: The protein and only RNP complex specifically binds to RNA and DNA.

Chapter 3:

3.Perturbation of temperature, salt and osmolytes on binding mechanism of Cas9:

3.1. Introduction:

The CRISPR/Cas9 system has emerged as a widely employed potent tool in biomedical research for a broad range of applications, with remarkable prospectives for genome editing alongside with significant potential in the imaging to track the desired genomic loci in realtime and point-of-care diagnostics. In order to achieve a precise thermodynamic information about Cas9's mechanism, we develop and standardize few biophysical experiments for Cas9-based genome editing to predict how changing various physiologicalconditions like temperature and pH as well as chemical changes like salt concentration

influence the physical phenomena of dCas9-based binding and Cas9-based cleavage at a certain DNA site. Here, solution state Circular Dichroism (CD) experiments and UV (Ultra-Violet) melting methods are used to capture the structure and stability of Cas9. Interestingly, binding study using Microscale Thrmophoresis (MST) reveals the RNA interaction with Cas9 is proved to be enthalpy driven whereas RNA guided Cas9 targetingto DNA is enthalpically deficit process explaining the energetics of Cas9 mechanism. With an improved understanding of Cas9's energetics, we propose that the extention of this study to DNA off target can boost to build up a tool for sgRNA designing and DNA off targeting based on thermodynamic parameters.

Cas9 is an endonuclease (~158 kDa) with a bi-lobed construction consisting of α-helical recognition (REC) and nuclease (NUC) lobes. The REC lobe comprises of mainly three regions namely bridge helix, REC1 domain, and REC2 domain. REC1 adopts an elongated, α-helical structure containing α helices and β sheets, whereas

REC2 adopts a helix containing bundle-like structure. The NUC lobe contains the PAM-interacting (PI) domain,which forms an elongated structure comprising of α helices and β sheets. NUC lobe is composed of two separate domains, RUVC and HNH alongwith a C-terminal domain (CTD) which interacts with the PAM duplex DNA. Though structural studies of Cas9 using X-ray crystallography and cryo-EM unveil domain architecture of Cas9, single-molecule Förster resonance energy transfer (sm-FRET) techniques capture conformational dynamicsand kinetics of Cas9 during binding and cleavage of DNA substrate, thermodynamic details of Cas9 and sgRNA binding followed by target DNA interrogation by Cas9 RNP is unclear. Here, we first report formation of Cas9 ternary complex is enthalpically deficit denoting the large positive change in enthalpy.

Despite significant strides in mechanistic understanding, of Cas9 function, few fundamentals in terms of thermal stability of Cas9, its salt-dependent structural changes, Cas9 activity (binding and cleavage) as a function of salt concentration and temperature is utterly lacking. Recent research demonstrated that cultured cells were much more effective at enhancing the effectiveness of genome editing by transcription activator-like effector nuclease (TALEN) and zinc finger nuclease (ZFN) mediated editing when maintained in transitory hypothermic (30°C) conditions. It was also demonstrated that, although greatly improving ZFN and TALEN performance, hypothermia therapy significantly decreases the effectiveness of CRISPR-Cas9 genome editing.

In this study, we investigated how the structure and stability of SpCas9 changes in monovalent salt concentration dependent and temperature-dependent manner. Then we asked how salt concentration and temperature perturb the binding events of Cas9 and sgRNA followed by Cas9 RNP and DNA target. Interestingly, analysis of the temperaturechanges on the binding events reveal the thermodynamic parameters of interaction therebygiving a clear understanding of the physical properties including mode and probability of occurrence of the interaction. The significance of this study also has its implications on SpCas9 applications in a broad spectrum of species of which many do not live at 37°C likezebrafish 28°C, C. elegans at 15-25°C and others. By observing the effect of temperature and salt concentration, it will help us to decide the range of temperature and salt concentration mediated changes in structure and function of CRISPR-Cas9 system.

3.1.1. Solution-state structure and thermal stability of Cas9:

To solve the solution state structure of SpCas9, we have used 20 mM HEPES-KOH, pH 7.5 buffer containing 150 mM KCl, 10mM $MgCl_2$, 1mM DTT and 10% glycerol which isin line with the buffer conditions used in other reports for in-vitro studies, unless stated otherwise. To investigate the secondary structural arrangement and stability of SpCas9 weused far UV region of Circular dichroism and found two negative minimum at 208nm and 222 nm and, a positive maximum at 190nm which shows alpha conformation while negative minimum at 218nm shows beta sheet structure. Analysis of the far-UV CD data showed 66% α-helical and 5% β-sheet conformation (Table 3.1.1.) which corroborates well with secondary structure obtained from the crystal structure of Cas9 (PDB ID: 4OO8) reported earlier. No CD signal was observed above 250nm indicating that the contributionof tyrosine, phenylalanine and tryptophan residues to the overall CD differential absorption are negligible. Temperature dependent CD spectroscopy is used to deduce the alteration ofsecondary structures of SpCas9 as a function of temperature. To test the thermal stability of SpCas9, far-ultraviolet CD spectra were recorded over a temperature range from 20 to 60°C (Supp Fig 3.1.1.a). The temperature downscan shows no change in ellipticity at any wavelength till 41°C indicating retention of helical conformation. Conformational transition starts occurring from 41°C and persists till 49°C before undergoing denaturationas indicated by spectral conformational changes of peaks at 208, 222 and 218nm characteristic of a mixture of short polypeptides. To decipher the denaturation temperatureof Cas9, CD melting experiment was performed and the melting curve (θ vs Temperature)resulted in a relatively sharp sigmoidal transition which signifies protein unfolding in a two-state cooperative fashion indicating a melting temperature of 43.0 (±0.5)°C (Fig 3.1.1.b) consistent with previous reports. When Cas9 unfolds, it loses its highly ordered structure and this is reflected in the CD spectral changes, thereby indicating the unfoldingof protein. To check whether Cas9 can restore back its secondary structure, the reverse cooling cycle was performed using UV spectroscopy with -0.2°C/min ramp rate which showed that Cas9 has irreversibly lost its structure during the heating cycle (Supporting Fig 3.1.1b). This irreversible unfolding of Cas9 is mainly due to the aggregation of protein at higher temperature.

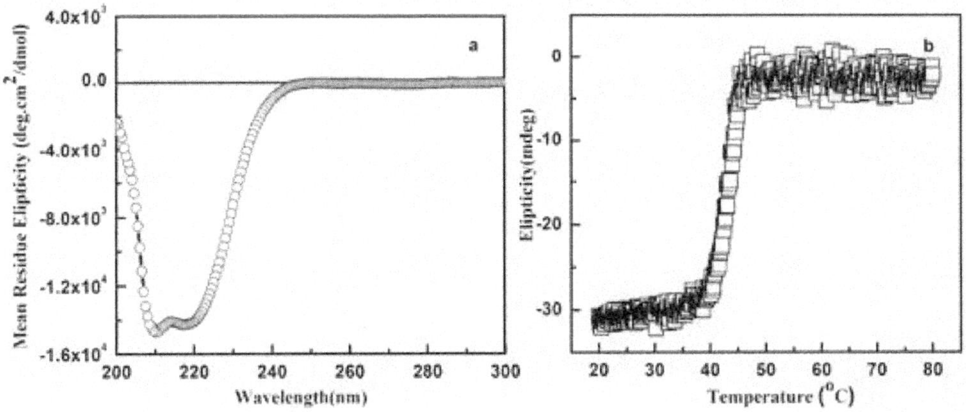

Fig 3.1.1. a) CD spectra of 0.6mg/ml WTSpCas9 in 20mM HEPES pH 7.5, 150mM KCl,1mM DTT and 10% glycerol at 20 °C.

b) Melting of 0.6mg/ml WTSpCas9 from 20 °C to 80 °C in 20mM HEPES pH 7.5, 150mM KCl, 1mM DTT and 10% glycerol at 1 °C/min ramp rate using CD spectroscopy.

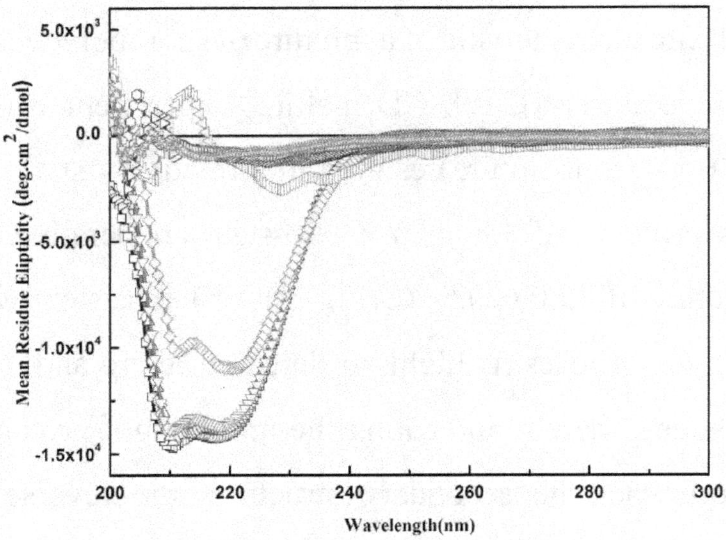

Supporting Information Figure 3.1.1.a:CD spectra of 0.6mg/ml Cas9 at different temperatures from 20 °C to 60 °C at each 5 °C temperature interval in 20mM HEPES pH 7.5, 150mM KCl, 1mM DTT, 10% glycerol.

Colour code- 20 °c black; 25 °c red; 30 °c blue; 35 °c olive; 40 °c magenta; 45 °c dark yellow;50 °c royal blue; 55 °c wine; 60 °c pink

Table 3.1.1.: Temperature dependent structural change of Cas9 by CD spectroscopy

Temperature (^{O}C)	% α helix	% of β sheet	Random Coil	Turn
20	66	5	29	0
25	66	5	29	0
30	63	6	31	0
35	62	6	31	0
40	46	22	32	0
45	4	48	48	0
50	10	43	47	0
55	4	48	48	0
60	5	47	48	0

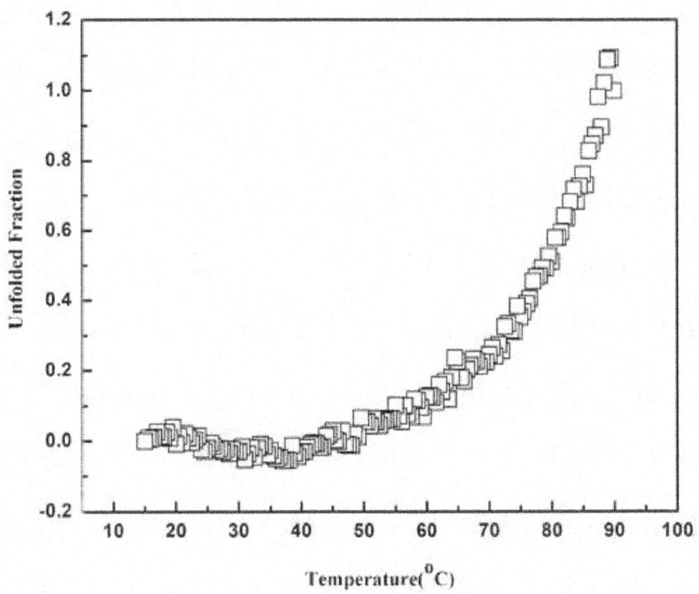

Supporting Information Figure 3.1.1 b:

Cooling curve for 0.03mg/ml SpCas9 in 20Mm HEPES pH 7.5, 150mM KCl, 1mM DTT,10% glycerol buffer with -0.2 °C/min ramp rate at 280nm wavelength using UV spectrophotometer.

3.1.2. Microscale Thermophoresis binding assay

With recent advancement in the technology of the binding affinity measurement, Microscale Thermophoresis (MST) has become an important tool for the direct measurement of equilibrium dissociation constant in various biological interactions. All the thermodynamic parameters such as Gibbs free energy change (ΔG), enthalpy change (ΔH), and entropy change (ΔS) can be calculated from the binding constant value of a temperature dependent binding experiment. GFP fused dSpCas9 was used for the binding experiment. In addition to IVT sgRNAs that were 12% urea PAGE purified, dSpCas9-GFP protein was employed. IVT sgRNAs were, it should be mentioned, purified by 12% urea-PAGE. The binding affinities of 180nM Cas9 protein and sgRNA with altered concentration (varying from 0.45nM to 15 μM) incubated at 25 °C for 10 mins at different KCl concentration starting from 82.5 mM to 125 mM and temperature ranging from 22 °C to 40 °C in the reaction buffer (20 mM HEPES, pH7.5, 150mM KCl, 1mM DTT, 10mM MgC12) were calculated using Monolith NT. 115 (NanoTemper Technologies GmbH, Munich, Germany). For the second step of binding between RNP complex and target DNA RNP complex (Cas9:sgRNA molar ratio, 1:1) was reconstituted in reaction buffer at 25 °C for 10 minutes in order to bind to target DNA for the second step (20 mM HEPES, pH7.5, 150mM KCl, 1mM DTT, 10mM MgC12). RNP complex was incubated with 30 bp dsDNA (IDT) duplex of several genomic loci that had undergone HPLC purification at variable doses (range from 0.9 nM to 30 mM) for 30 min in reaction buffer at various temperatures (22 °C to 40 °C). The sample was placed into the Nanotemper-provided standard-treated capillaries. In order to conduct the measurements, blue filters with LED powers of 20%, 40%, and 80% were used. These filters have emission wavelengths of 500–550 nm and excitation wavelengths of 465–490 nm, respectively. For every measurement, every experiment was run at least twice. Software called NanoTemper was used for data analysis. For extracting the thermodynamic parameters of temperature dependent binding study betweenCas9 and sgRNA the equations used were-

$$\Delta G^o = -RT \ln K_a \quad (1)$$

$$K_d = 1/K_a \quad (2)$$

$$\Delta G^o = \Delta H^o - T\,\Delta S^o; \quad (3)$$

$$-RT\ln K_a = \Delta H^o - T\,\Delta S^o \quad (4)$$

$$\ln K_a = -\Delta H^o/RT + \Delta S^o/R \quad (5)$$

Where K_a is equilibrium binding association constant; ΔG^o is change in gibbs' free energy;K_d is equilibrium binding dissociation constant; ΔH^o is change in enthalpy; ΔS^o change in entropy. Equation (5) is the equation of a straight line with a slope of -ΔH^o/R and an intercept equal to ΔS^o/R. To determine ΔH^o and ΔS^o, lnK_a vs 1/T was plotted. The generated straight line will have a slope equal to -ΔH^o/R and an intercept equal to ΔS^o/R.

To extract the contribution of electrostatic energy salt dependent binding study was performed between Cas9 and sgRNA followed by DNA and RNP complex the equations used were-

$$\Delta G_{obs} = \Delta G_{pe} + \Delta G_t \quad (6)$$

ΔG_{pe} is electrostatic contribution and ΔG_t is the non-polyelectrolyte contribution. ΔG_{obs} is the overall observed Gibb's free energy.

$$\Delta G_{pe} = [SK]\ RT\ \ln\ [Salt] \qquad (7)$$

$$\Delta G_{obs} = -RT\ \ln K_{obs} \qquad (8)$$

[Salt] is the concentration of KCl at which the experiment was performed. SK denotes theno of salt counterion involved in the process. From the plot of $\ln K_a$ vs ln [Salt], the slope, (-SK) was calculated using equation (1) and (7).

3.2. Perturbation of Temperature on binding mechanism of Cas9:

3.2.1.Effect of temperature on sgRNA binding by Cas9:

To examine the thermodynamics of the Cas9-sgRNA binding, we performed a binding experiment using Microscale Thermophoresis which is sensitive to give the value of thermodynamic equilibrium constant associated with Cas9 binding to sgRNA. To this end, we used recombinant catalytically *dead* (inactive) SpCas9 (dSpCas9) tagged with green fluorescent protein (GFP) previously reported. We observed that at 22°C the binding affinity is 8.5×10^6 M^{-1} whereas at 40°C the binding association constant comes out to be 3.16×10^5 M^{-1} (Figure 3.2.1a). Notably, we did not go beyond 40°C as the denaturation temperature of protein is ~43.0°C. The thermodynamic parameters were calculated as ΔH to be -139.8 $KJmol^{-1}$ and ΔS to be -0.34 $KJmol^{-1}$ from the slope and intercept of the linearplot (Fig 3.2.1b) using equation (5) (materials and methods). The overall ΔG of binding resulted in -38.5 KJ mol^{-1} at 25°C suggesting that the process of binding is spontaneous. Briefly, with varying temperature, the binding affinity changes during the formation of RNP complex which denotes that this is an exothermic process with negative entropy change resulting in an overall negative gibbs' free energy change. Thermodynamic parameters (Table 3.2.1) at 25°C resulted in a negative entropy change in the pretarget (i.e.Cas9-sgRNA binary complex), before interaction with target in agreement with the fact that PAM-interacting domain and seed sequence (i.e. crRNA) from guide RNA are pre- organized and more ordered for PAM recognition in active Cas9 than the largely disordered PAM interacting domain of inactive Cas9. Similarly, using single molecule FRET (smFRET) study it was observed that sgRNA interacts with Cas9 to form RNP complex and during this process inactive Cas9 gets activated by changing into a rigid conformation(ordered state) from a flexible mode of action (disordered state) in accordance with the negative entropy change observed in our study. The association of sgRNA with Cas9 exhibited negative changes in the binding enthalpies indicating sufficient amount of heat release during the interaction. During sgRNA binding to Cas9, sequence dependent interaction of tracrRNA followed by sequence independent interaction of crRNA with Cas9 domains take place forming salt bridges, H bonds and Van der Waal's interaction along with an extent of ionic interaction and thereby releasing large amount of heat suggesting this binding to be enthalpically driven. Also, this entropically deficient and enthalpically driven process couples with the process of change in major structural rearrangements i.e. conversion into a more ordered state, with the movement of REC lobe towards the HNH domain suggesting an overall energetically rich process (with negative enthalpy).

Table 3.2.1: Binding parameters obtained from MST experiments in 20mM HEPES pH7.5, 150mM KCl, 1mM DTT, 10mM $MgCl_2$ at different temperatures.

Binding of Cas9 and sgRNA		**Binding of RNP and DNA target**	
Temperature(°C)	**K_{d1} (M)**	**Temperature(°C)**	**K_{d2} (M)**
22	**1.2×10^{-7}**	**25**	**7.4×10^{-7}**
25	**3.2×10^{-7}**	**30**	**4.4×10^{-7}**
30	**5.8×10^{-7}**	**37**	**5.6×10^{-8}**
35	**2.1×10^{-6}**		
40	**3.2×10^{-6}**		

K_{d1} is the equilibrium dissociation constant of sgRNA for the first step of binding site to Cas9. K_{d2} is the equilibrium dissociation constant of DNA target for the second step of binding to RNP complex.

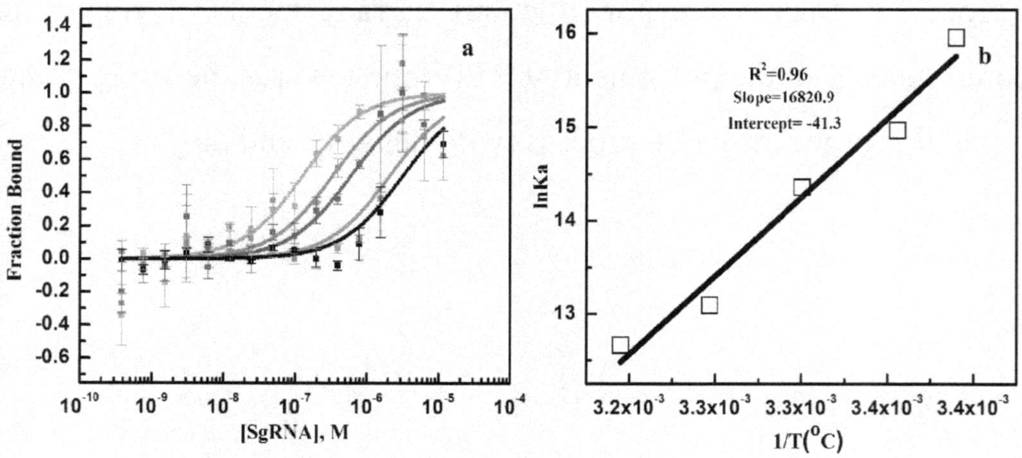

Figure 3.2.1. a) MST results showing binding of dSpCas9-GFP protein to sgRNA against*VEGFA3* DNA target expressed as fraction bound units (y-axis) with respect to varying concentrations of purified sgRNA (x-axis) at different temperatures varying from 22°Cto 45°C in the assay buffer containing 20 mM HEPES of pH7.5, 150mM KCl, 1mM DTT,10mM $MgCl_2$. The concentration of sgRNA in the titration was varied from 1.5×10^{-5} M (upper point) to 4.5×10^{-10} M (lower point). The sgRNA sequence is indicated. Error bars represent SD (2 independent experiments).

Colour code: 22 °C - Magenta; 25 °C - Olive; 30 °C - Blue; 35 °C - Red; 40 °C - Black.

b) Plot of $\ln K_a$ (ln of binding association constant) vs $1/T$ (1 / specific temperature at which experiment was performed) for the binding studies of protein (dSp-GFP) and sgRNA at varied temperatures. The resulting data was analyzed by linear regression method using equation (1) and (5) from the materials and methods.

3.2.2. Effect of temperature on DNA Binding by Cas9·sgRNA:

Next, to examine the binding affinity of the DNA and RNP complex, MST experiment was carried out under varied temperatures. We noticed that with increasing temperature, the binding affinity of DNA target to RNP complex increases as the binding dissociation constant decreases (Equation 2, materials and methods). At 25 °C, the binding associationconstant is 1.63×10^6 M^{-1} whereas at 37°C, the binding affinity of the system is 1.74×10^7 which is one order higher than that at 25 °C (Fig 3.2.2a). From the slope and intercept of $\ln K_a$ vs $1/T$ plot (Fig 3.2.2b) ΔH, ΔS and ΔG parameters were calculated to be 168.2 $KJmol^{-1}$; 0.62 $KJmol^{-1}$ and -16.8 $KJmol^{-1}$ respectively, at 25°C, which rules out that the binding event is an endothermic, entropy driven and spontaneous process (Table 3.2.2). The activity of Cas9 depends on its interaction to DNA substrate through RNP complex formation andoccurs in three sub-steps involving PAM binding, target DNA melting and DNA:RNA heteroduplex formation. We assume that the first substep (hydrogen bonding of Cas9's PIdomain to PAM sequence in DNA target) and the third sub-step (the generation of DNA :RNA heteroduplex) are enthalpy driven processes, showing a negative entropy change dueto conversion from a disordered to an ordered state, whereas the second step (unwinding of dsDNA) is highly endothermic, being a random entropy driven process, thereby makingthe overall sum of the three steps to have a highly positive value both in terms of enthalpyand entropy change. This

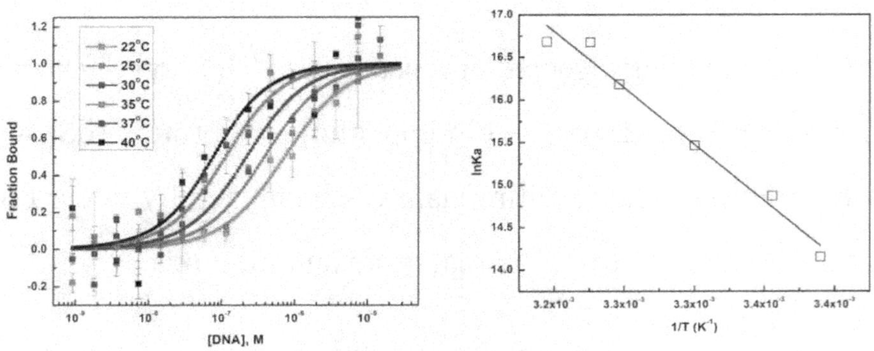

supports the fact that genome editing is an entropy driven process.

Figure 3.2.2: a) Binding of RNP complex and *VEGFA3* DNA target at varied temperature(22, 25, 30, 35, 37 and 40°C) in the assay buffer containing 20 mM HEPES of pH7.5, 150mM KCl, 1mM DTT, 10mM $MgCl_2$ using MST. DNA was titrated from 2.5×10^{-5} M to 9.0×10^{-11} M. Sequence of target DNA is shown. Error bars represent SD (n=2).

Colour code: 22 °C - magenta, 25 °C - Olive; 30 °C - Blue; 35 °C – Red, 37 °C -Brown, 40 °C -Black.

b) For the binding study at different temperatures of RNP complex and Target DNA, plot of lnK_a vs 1/T is used to calculate the thermodynamic parameters of binding. The resultingdata was analyzed by linear regression method using equation (1) and (5) from the materials and methods.

Table 3.2.2: Thermodynamic parameters obtained from MST experiments conducted in20mM HEPES pH 7.5, 150 mM KCl, 1 mM DTT, 10 mM $MgCl_2$ at 25°C.

Binding Events	ΔH ($KJmol^{-1}$)	ΔS($KJmol^{-1}K^{-1}$)	ΔG($kJmol^{-1}$) at 298K
Cas9-sgRNA	**-139.8**	**-0.34**	**-38.5**
RNP Complex-DNA target	**168.2**	**0.62**	**-16.8**

3.3 Effect of salts on binding interaction of Cas9:

3.3.1 Cas9 structure and stability depends on optimum salt concentration:

As the presence of different inorganic ions influence the alpha helices and beta sheets of protein based on Hofmeister series we wanted to see the effect of monovalent salt (KCl) on the secondary structure of Cas9 using CD spectroscopy. The far-UV CD spectrum in the wavelength range 190-250nm was captured where the ellipticity at two negative minima at 208nm and 222nm and a positive maximum at 190nm increased slightly with increasingKCl concentration indicates almost no change in percentage of alpha helix contribution tothe overall secondary structure (Supporting Fig 3.3.1a). The mean residue ellipticity of Cas9 at 222nm wavelength has increased to a lesser extent with increasing KCl concentrations and analysis of the far-UV CD data resulted in 19.0% α-helical and 25.2% beta-sheet structure at 0mM KCl which becomes 16.2% α and 18.4% beta structure at 1000mM

KCl, depicting 0.08% reduction in α helical and 15.6% decrease in beta sheet contribution in SpCas9 (Fig 3.3.1a, Supporting Fig 3.3.1a).

To test the stability of protein in presence of varied salt concentration, salt dependent thermal melting (T_m) of Cas9 was studied (Supp Information 3.3.1b). We observed that Cas9 unfolds irreversibly by a two-state process and its thermal stability is sensitive to saltconcentration. The melting temperature (T_m) as a function of KCl concentration indicates a sharp increase of T_m on the ion molarity suggesting protein's instability at 0mM KCl andgradual increase of stability upto 500mM KCl (Fig 2B). At 500mM KCl, T_m is 43.7°C whereas at 1000mM KCl, the melting temperature reaches its saturation. The observation illustrates that increment of ions and, therefore, electrostatic interaction of ions on Cas9's surface is proposed to enhance the protein's thermal stability due to the nonspecific interactions between the solvated ions and the polar regions of the Cas9 backbone in accordance to a previous study. Monovalent ions present in KCl increase the surface tension of the solution, which results in stabilization of the folded conformation of Cas9. The outcome of this melting study is complementary to other systems in which charged residues stabilize the protein through a network of inter-residue electrostatic interactions.

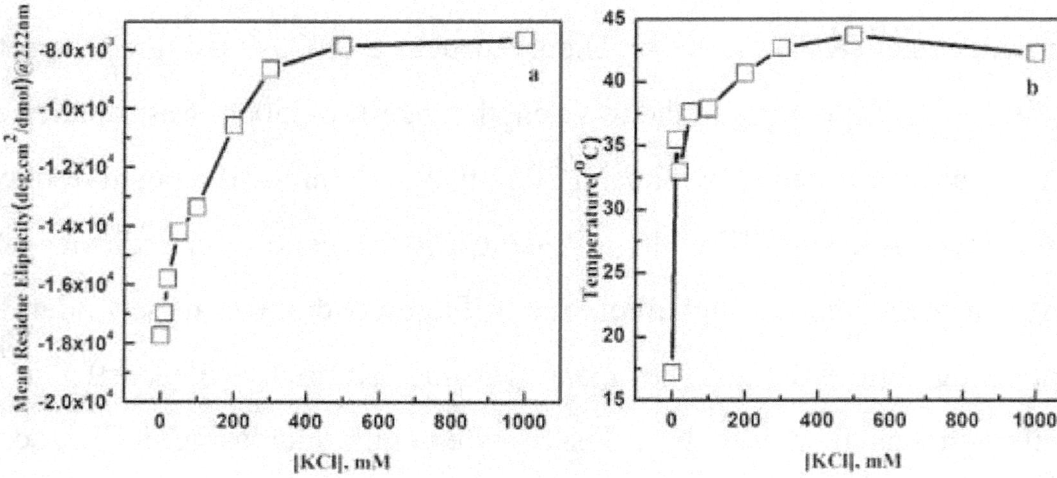

Figure 3.3.1: a) Mean residue ellipticity values at 222 nm of 0.25mg/ml WTSpCas9

at different KCl concentrations ranging from 0 mM to 1000 mM in 20mM HEPES buffer (pH7.5) at 25°C.

b) Plot of melting temperature (T_m) versus salt (KCl) concentration for the UV-melting of

0.03 mg/ml WTSpCas9 with 0.2°C/min ramp rate in a buffer containing 20mM HEPES (pH7.5) with varied salt concentration starting from 0mM to 1000mM. The experiment was done at 280nm fixed wavelength.

Table 3.3.1: Salt dependent structural change of Cas9 by CD spectroscopy

Salt Concentration (mM)	**%α helix**	**%β sheet**	**Random coil**	**Turn**
0	**19**	**25.2**	**33.1**	**22.7**
10	**14.9**	**33**	**32.8**	**19.4**
20	**17.3**	**30.8**	**32.6**	**19.4**
50	**16.9**	**34.9**	**29.7**	**18.4**
100	**19.8**	**14**	**38.9**	**27.3**
200	**17.6**	**28.6**	**30.6**	**23.1**
300	**20**	**22.6**	**32.9**	**24.5**
500	**16.6**	**34.2**	**33.4**	**15.9**

1000	16.2	18.4	42.7	22.7

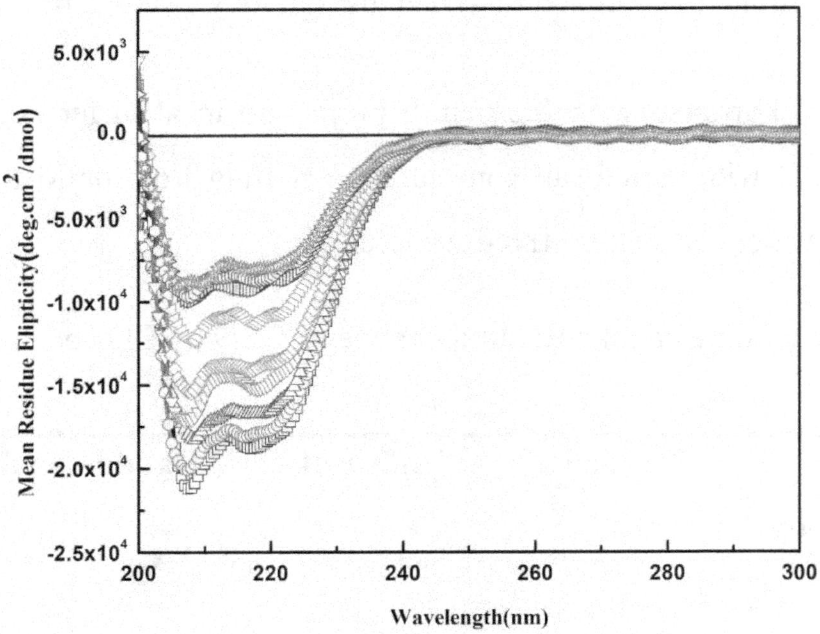

Supporting Information Fig 3.3.1a:

CD spectra of 0.25mg/ml Cas9 protein at different KCl concentrations ranging from 0 mM to 1000 mM in 20mM HEPES buffer (pH7.5) at 25°C represented in the wavelength rangeof 200nm to 300nm.

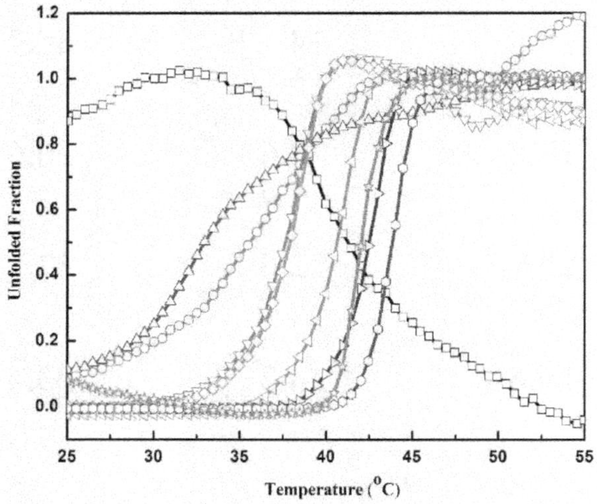

Supporting Information Fig 3.3.1b: UV melting experiment of 0.03mg/ml Cas9 proteinat 0.2°C/min ramp rate and at 280nm wavelength in a buffer containing 20mM HEPES (pH7.5) with varied salt concentration starting from 0mM to 1000mM.

3.3.2. Influence of salt concentration on the binding mechanism of Cas9 and sgRNA:

For investigating the role of salt on the binding of Cas9-sgRNA, we first checked the binding affinity of sgRNA with Cas9 protein for RNP complex formation at different salt concentrations using MST. Complex formation carried in the presence of increasing salt concentration showed a decrease in binding dissociation constant (K_d) value; which meansincrease in association constant (K_a) value (Equation 2, materials and methods). At 82.5 mM KCl concentration the binding equilibrium dissociation constant (K_d) is 1.87×10^{-7} M whereas at 125 mM KCl the K_d is 5.01×10^{-8} M (Fig 3.3.2a) indicating the gradual decrease of K_d with variation of KCl (Table 3.3.2). According to polyelectrolyte theory[41] the free energy parameter (ΔG_{obs}) for binding between Cas9 and sgRNA can be dissected into electrostatic (ΔG_{pe}) and non-polyelectrolytic (ΔG_t) terms using equation (6) (materials and methods). The electrostatic contribution to ΔG_{obs} is largely due to coupled polyelectrolyteeffects, the most important of which is the net release of condensed counterions while the non-polyelectrolyte contribution is due to hydrogen bonding, Van der Waals and hydrophobic interactions upon RNP complex formation during complexation of Cas9 withsgRNA. In order to estimate the contribution from the ΔG_{pe}, we monitored the change in binding affinity of Cas9 to sgRNA as a function of increasing salt concentration. The slopeof lnK_a versus ln $[K^+]$ was used to obtain the effect on the electrostatic interaction upon RNP complex formation (Figure 3.3.2b). Noteworthy was the influence of modification onthe electrostatic interactions, that contributed

15.1 KJmol^{-1} (non favourable process,positive quantity) (Materials and methods: equation 7) while the overall contribution of ΔG_{obs} is -40.6 KJmol^{-1} (favourable process, negative quantity) (Materials and methods: equation 8) at 112.5mM KCl (Table 3.3.2b), suggesting that conformational change of Cas9 is favoured by primarily affecting the non-electrolytic interactions which contributesthe fullest to the overall interaction. Monitoring the change in binding affinity with varied salt concentration allowed an estimation of the electrostatic component to the observed binding free energy and indicated that sgRNA binding to Cas9 results predominantly in thenon-electrostatic interactions upon RNP complex formation. In agreement with the previous report, in the process of recognition by Cas9, guide region of sgRNA (i.e. crRNA)utlilizes majorly its backbone of phosphate groups to interact with the REC1 and bridge helix domain whereas few 2' hydroxyl groups of sgRNA form hydrogen bonds with Valine, Threonine, Tyrosine and Arginine of Cas9 domain thereby contributing non- electrostatic component favourably to overall free energy. Our finding of almost no electrostatic contribution is consistent with the fact that only three arginine residues (Arg66, Arg70, and Arg74) of Cas9 form multiple salt bridges with the sgRNA backbone, whereas Arg78 andArg165 form a single salt bridge contributing very low energy to the overall electrostatic interaction. Again, Favourable non-electrostatic contribution is visualized by the events inwhich non-ionic interactions or water release play a severe role over ionic interactions andion release which lies linearly with the previous observation of very low salt counterion release (Table 3.3.2b).

Table 3.3.2: Binding parameters obtained from MST experiments in 20mM HEPES pH 7.5, 1mM DTT and at different salt concentration.

Binding of Cas9 and sgRNA		**Binding of RNP and DNA target**	
[Salt], mM	K_{d3} **(M)**	**[Salt],mM**	K_{d4} **(M)**
82.5	1.87×10^{-7}	**10**	2.7×10^{-5}
87.5	1.25×10^{-7}	**40**	2.6×10^{-7}
100	1.16×10^{-7}	**45**	1.3×10^{-7}
112.5	7.64×10^{-8}	**62.5**	7.4×10^{-8}

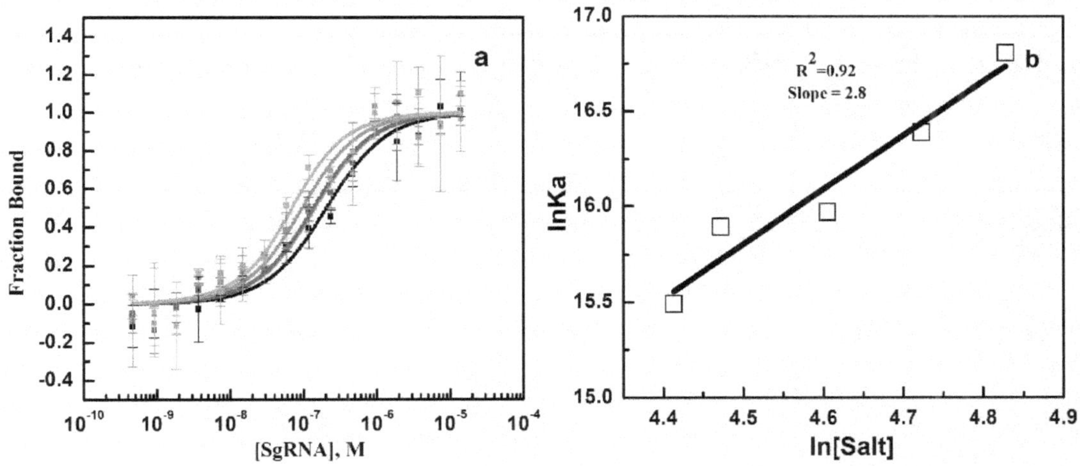

Figure 3.3.2a: a) MST results showing binding of dSpCas9-GFP protein to sgRNA against *VEGFA3* DNA target expressed as fraction bound units (y-axis) with respect to varying concentrations of purified sgRNA (x-axis) at different KCl concentrations (82.5, 87.5, 100,

112.5 and 125mM) in the assay buffer containing 20 mM HEPES of pH7.5 and 1mM DTT.The concentration of sgRNA in the titration was varied from 1.5×10^{-5} M (upper point) to 4.5×10^{-10} M (lower point). The sgRNA sequence is indicated. Error bars represent SD (2 independent experiments).

Colour code: 82.5mM KCl-black; 87.5mM KCl-red; 100mM KCl-blue; 112.5mM KCl- olive; 125mM KCl- magenta.

b) Plot of lnK_a (ln of binding association constant) vs ln [Salt] for the binding studies of protein (dSp-GFP) and sgRNA at varied salt concentration. For the binding study at different monovalent salt concentration of Cas9 and sgRNA, plot of lnK_a vs ln [Salt] is used to calculate the slope (–SK), where SK denotes the no of salt counterion involved in the interaction. The resulting data was analyzed by linear regression method using equation
(7) and (8) from the materials and methods.

Table 3.3.2 b: Contribution of electrostatic component on Complex formation.

At 112.5 mM [Salt]	Sk	(Δ*G*pe)*(kJmol^{-1})	Δ*G*obs (kJmol^{-1}) at 298K
Cas9-sgRNA	**-2.8**	**15.1**	**-40.6**
RNP complex-DNA target	**-3.1**	**16.8**	**-44.5**

3.3.3.DNA binding of Cas9·sgRNA are affected by salt concentration:

For determining the effect of monovalent salt concentration on the physical propertiesof ternary complex (RNP complex : DNA) formation, binding experiments were performed. MST technique was used to perform the binding experiment between DNA andRNP complex in presence of varying concentrations of KCl at 25°C. Ternary complex formation carried in the presence of increasing salt concentration showed an decrease in binding dissociation constant (K_d) value; which means increase in association constant (K_a) (Equation 2, materials and methods). Binding experiments was carried out in presence of 10 mM KCl and revealed a ~1600-fold less binding affinity of DNA interrogation (k_d=10^{-5} M) by Cas9·sgRNA than at 112.5 mM KCl (k_d =10^{-8} M) (Figure 3.3.3a). (Table 3.3.2) Accordingly, the monovalent ion accelerates the assembly of the ternary complex. However, the overall stability of the ternary complex was affected as the K_d for DNA binding appreciably improved (K_d with 112.5 mM KCl = 1.6×10^{-8} M; K_d with 10 mM KCl = 2.7×10^{-5} M) with presence of salt (Table 3.3.2).

The interpretation of lnK_a vs ln [Salt] (Fig 3.3.3b) was the influence of salt concentration on the electrostatic interactions (ΔG_{pe}) which contributed 16.8 KJmol^{-1} (non favourable process, positive quantity) (Materials and methods: equation 7) while the overall contribution of ΔG_{obs} is -44.5 KJmol^{-1} (favourable process, negative quantity) (Table 3.3.2b;Materials and methods: equation 8), suggesting that the generation of heteroduplexduring ternary complex formation is favoured by primarily affecting the non-electrolytic interactions. This observation of fully favourable non-electrolytic contributiion is in accordance with a previous report explaining the interaction of DNA backbone phosphategroups with REC1, RUVC and PI doamins, generation of Van der Waals interactions between C2' atoms of target DNA and REC1, RUVC domain of Cas9 contributing towardsthe ability of Cas9 to discriminate between DNA and RNA targets, and finally formation of Watson-Crick base pairing between DNA target and 20 nt long seed region from sgRNAduring generation of R loop. This observation leads to conclude that at a very high salt concentration, the activity diminishes almost due to the suppression of ionic interaction. Due to the high charges on fully ionized Cas9, sgRNA, and DNA, ionic and non-ionic effects will contribute substantially to the stability of both specific and non-specific complexes. According to our observation, in higher salt concentrations non-specific interactions contribute at a large extent indicating lesser cleavage efficiency at higher salt concentration. To investigate the salt effect on the specific interaction during ternary complex formation we performed the binding experiment (EMSA) in presence of differentdivalent salt, $MgCl_2$. The interaction between RNA and protein produces an increase in the affinity of the specific 1:1 complex upon increasing the salt concentration, whereas very high salt concentration hampers the second step of binding i.e. formation of heteroduplex between RNA and DNA followed by activity too. Higher salt concentration (more than 200mM) weakens Cas9's activity by surrounding the negatively charged nucleic acids withpositive counterions (K^+) of salt, thereby creating an ionic crowding effect in the solution environment which is in line with a previous report. Interestingly, ternary complex formation is monitored by the optimum salt concentration i.e. in the range of 100-150 mMin concordance with the binding experiment (Fig 3.3.3a).

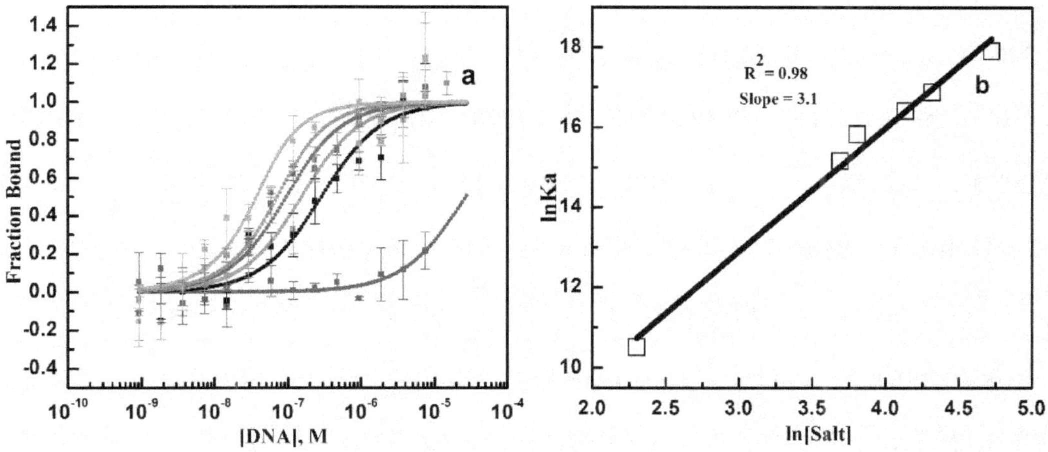

Figure 3.3.3: a) Binding of RNP complex and *VEGFA3* DNA target at varied monovalentsalt concentration (10, 40, 45, 62.5, 75, 112.5mM) in the assay buffer containing 20 mM HEPES of pH7.5 and 1mM DTT using MST. DNA was titrated from 2.5×10^{-5} M to 9.0×10^{-11} M. Sequence of target DNA is shown . Error bars represent S.D. (n=3).

Colour code: 10mM KCl-wine; 40mM KCl-black; 45mM KCl-red; 62.5 mM KCl-blue; 75mM KCl-olive; 112.5mM KCl- magenta.

b) For the binding study at different monovalent salt concentrations of RNP complex and Target DNA, plot of lnK_a vs ln [Salt] is used to calculate the slope (–SK), where SK denotes the no of salt counterion involved in the interaction. The resulting data was analyzed by linear regression method using equation (7) and (8) from materials and methods.

3.4. Effect of cosolutes on binding interaction of Cas9:

RNA-guided endonucleases of the CRISPR-Cas9 system have revolutionized the field of targeted genome engineering. The discovery of the molecular mechanism of recognition and cleavage of target DNA and the synthesis of a chimeric single guide RNA by fusing crRNA and tracrRNA facilitated the ease of programming this system to target any sequence. Other than editing genes, the CRISPR-Cas9 system has been used to activate orinhibit gene functions, doing epigenetic modification, large scale genetic screening, generation of animal models and even in imaging the genome. Various aspects of the molecular mechanism of Cas9 are elucidated with the help of biochemical and structural studies. One such study is the *in vitro* cleavage (IVC) activity assay in which purified Cas9 protein is complexed with sgRNA. The resultant ribonucleoprotein (RNP) complex is treated with a substrate containing target site and the cleaved products are observed in agarose gel. However, the IVC activity assay is not always reflective of the *in vivo* cleavage efficiency of the same RNP complex. One of the reasons for this difference could be the inaccessibility of target site due to it being in a closed conformation (heterochromatin form) *in vivo*. Another important yet unexplored explanation for this difference could be 'molecular crowding effect'. Inside cells, biological macromolecules are present in a high concentration with a volumeoccupancy range of 5% to 40%. The difference that occurs in the biochemical activities ofmolecules in crowded conditions with respect to dilute conditions is due to two reasons: excluded volume effect and chemical effects. Excluded volume effect is the condition wherein the volume occupied by macromolecules remains unavailable to other molecules.When the concentration of such macromolecules is increased in a solution, the volume that remains for other solutes proportionately gets decreased. As a result of volume exclusion, the thermodynamic activity of a biomolecule might get altered. On the other hand, chemical effects are a result of attraction and repulsion between molecules and it might alter the reaction towards either a stabilizing form or a destabilizing one, thereby providing either an additive or an opposite contribution to the excluded volume effect.The *in vitro* experiments to study biochemical reactions take place in a relatively dilute concentration which does not reflect the actual *in vivo* reaction conditions. The crowdedmilieu has high viscosity with a chance occurrence of specific as well as non-specific interactions amongst the molecules. Therefore, the phenomenon of molecular crowding should be factored in while doing biochemical studies.

In order to understand the effect of molecular crowding on genome editing efficiency of theCRISPR-Cas9 system, we selected chemicals- as molecular crowding agents- which are inert, non-charged polymers of certain sizes that occupy space but do not interact with target proteins.

We performed our standardized binding assay with SpCas9, in the presence of Sucrose and Trehalose- as small sized molecular crowding agents (Fig. 3.4.1). And we observed with increasing Sucrose concentration, the binding affinity gradually increases. Then another small molecular crowder we selected i.e. Trehalose. To understand the binding affinity with increasing Trehalose concentration, we observed the binding affinity increases again in smaller extent (Fig. 3.4.3). Through this study, we can infer that molecular crowding can affect the CRISPR-Cas9 system's genome editing efficiency at any one or all of the following three levels:

a. Regulating Cas9 protein's structure and stability.

b. Modulating Cas9 protein and sgRNA complexation.

c. RNP complex and target site interaction.

To be able to put the CRISPR-Cas9 system to a therapeutic use, the knowledge of how thesystem would behave in a crowded environment present *in vivo* becomes essential. More extensive experiments would be required to find this out.

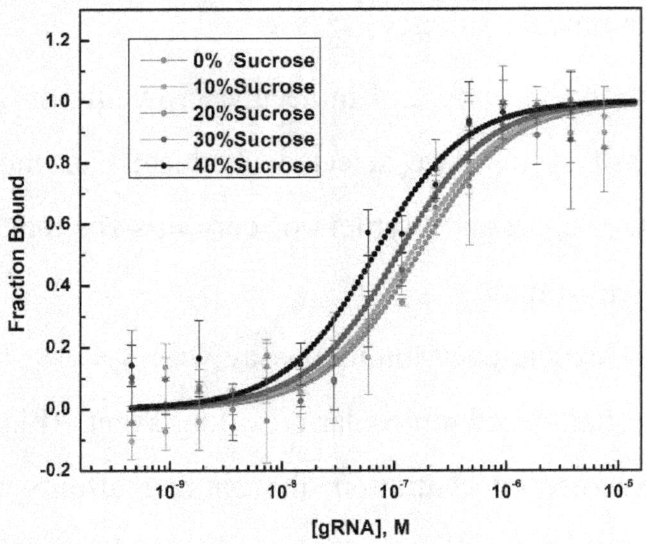

Figure3.4.1: In presence of sucrose binding between dSp-egfp and Vegfa3 RNA

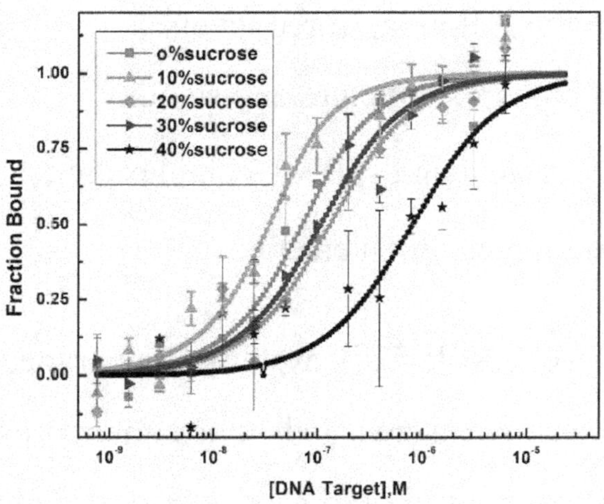

Figure3.4.2: In presence of sucrose binding between RNP complex made of dSp-egfpand Vegfa3 RNA and duplex DNA target

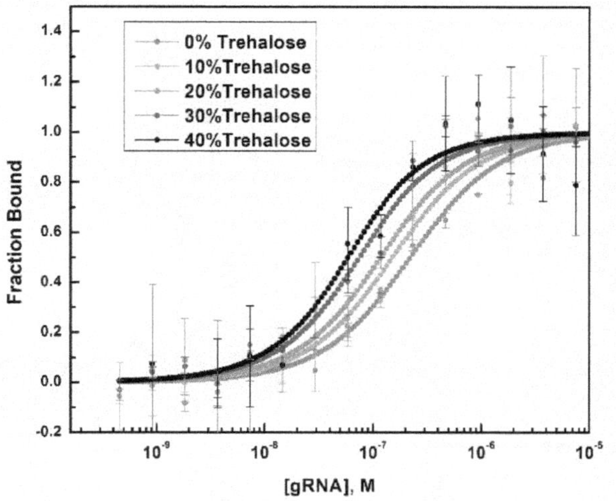

Figure 3.4.3: In presence of trehalose binding between dSp-egfp and Vegfa3 RNA

Solution-state structure and thermal stability of CjCas9

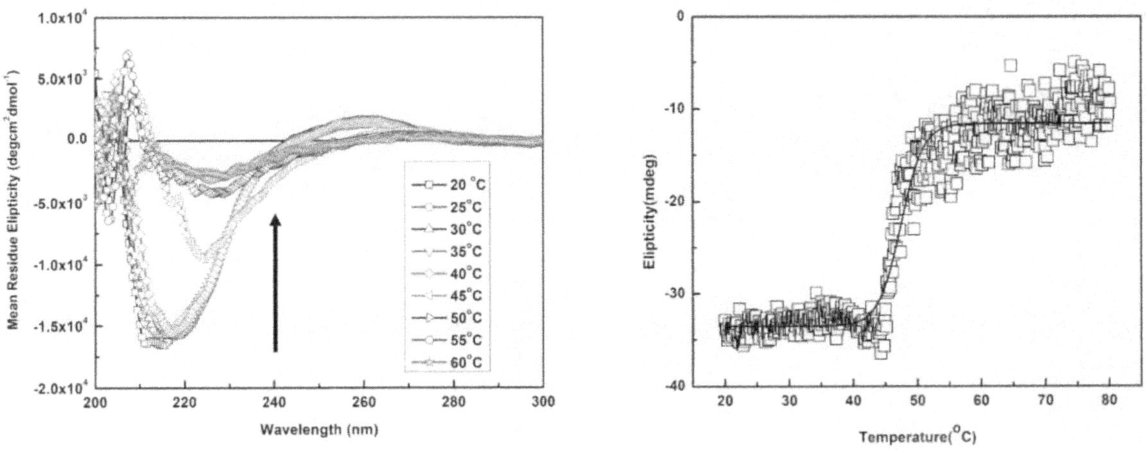

Tm=47.5 °C

Fig 3.4.4: Solution state structure and thermal stability of CjCas9

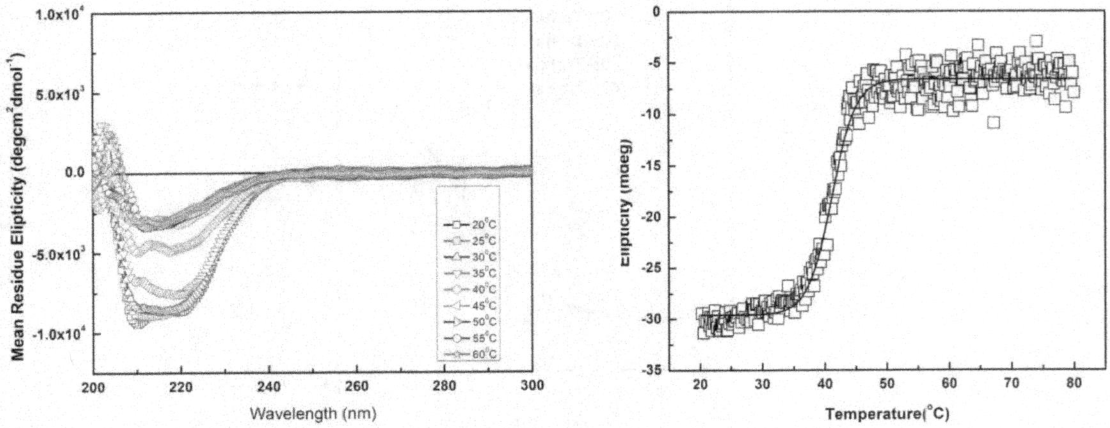

Fig 3.4.5: Solution state structure and thermal stability of SaCas9

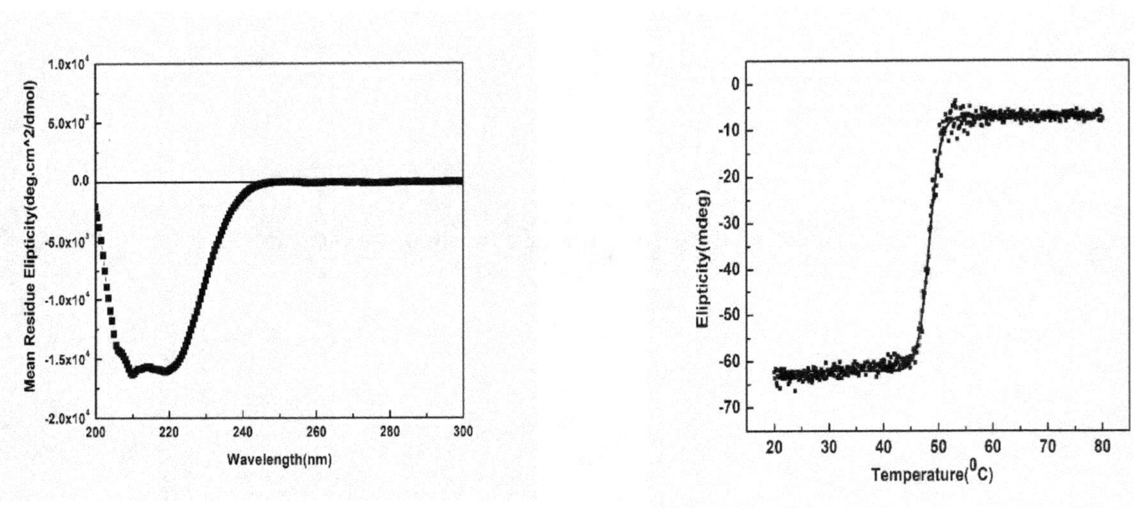

Fig 3.4.6: Solution state structure and thermal stability of WTFnCas9(N6)

3.4.1. Cas9 protein's stability using NanoDSF under different molecular crowdersand with varied salt concentration:

Protein's thermal stability got checked by using NanoDSF machine from Nanotemper. We performed several melting experiments using WTSpCas9 and WTFnCas9(N6) with various molecular crowders like small one- Sucrose, Trehalose, Ethylene glycol, and big one- PEG 200, PEG 4000, PEG 8000, Ficoll 400. (Table 3.4.1 a and 3.4.1 b) We also checked the Tm of WTSpCas9 protein under varied salt and glycerol concentration. And from the Tm of that we calculated several thermodynamic properties of protein like enthalpy, entropy and Gibbs' free energy change value.

peg200	**0%**	**41.7°C**
	10%	**40.6°C**
	20%	**39.1°C**
	30%	**38.9°C**
	40%	**37.9°C**
peg 4000	**0%**	**41.6°C**
	10%	**40.3°C**
	20%	**44.4°C**
	30%	**46.7°C**
	40%	**50.5°C**
peg 8000	**0%**	**41.7°C**
	10%	**42.2°C**
	20%	**33.2°C**
	30%	**28.5°C**
	40%	**28.1°C**
ficoll 400	**0%**	**41.7°C**

10%	**42.0°C**
20%	**41.5°C**
30%	**42.7°C**
40%	**38.3°C**
0 sucrose %	**41.7°C**
10%	**43.8°C**
20%	**45.5°C**
30%	**47.0°C**
40%	**49.1°C**
trehalose %	**41.5°C**
10%	**43.5°C**
20%	**45.4°C**
30%	**46.2°C**
40%	**49.1°C**
ethylene gluycol 0%	**41.6°C**
10%	**40.8°C**
20%	**39.1°C**
30%	**36.3°C**
40%	**33.0°C**

Table 3.4.1a: Tm of WTSpCas9 under different molecular Crowders.

PEG200 0%	47.0°C
10%	44.7°C
20%	42.3°C
30%	41.2°C
40%	37.8°C
PEG 4000 0%	47.2°C
10%	43.8°C
20%	45.7°C
30%	46.7°C
40%	48.7°C
PEG 8000 0%	47.0°C
10%	44.9°C
20%	40.8°C
30%	
40%	65.6°C
Ficoll 400 0%	47.1°C
10%	46.8°C
20%	46.5°C
30%	45.9°C
40%	
sucose 0%	47.2°C
10%	48.3°C
20%	49.6°C
30%	50.5°C
40%	51.3°C
trehalose 0%	47.2°C
10%	48.3°C
20%	49.7°C
30%	50.6°C
40%	51.0°C
Ethylene glycol 0%	47.0°C
10%	45.3°C
20%	43.0°C
30%	39.8°C
40%	35.4°C

Table 3.4.1 b: Tm of WTFnCas9 under different molecular Crowders.

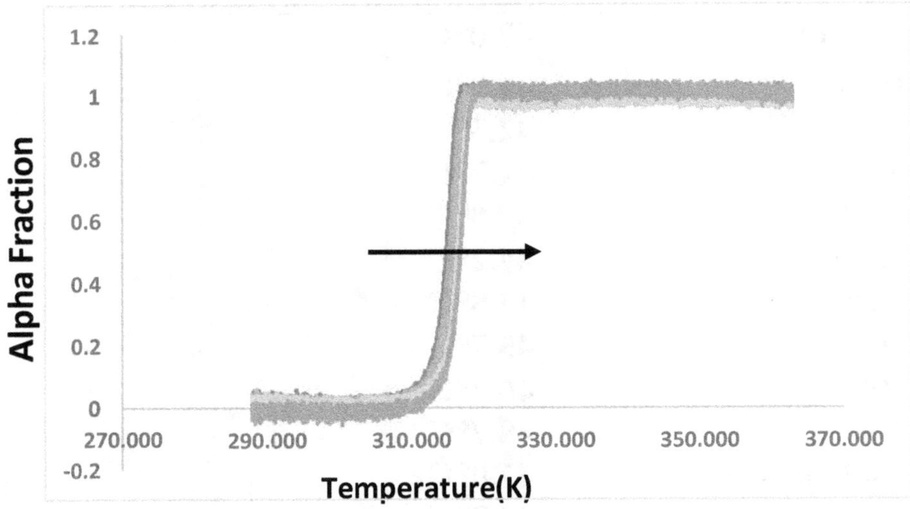

Figure 3.4.7: With increasing glycerol concentration, the melting temperature ofSpCas9 increases from 0% to 25% glycerol. Buffer Condition [Hepes-pH7.5(20mM),kCl(50mM),$MgCl_2$(10mM),DTT(1mM)]

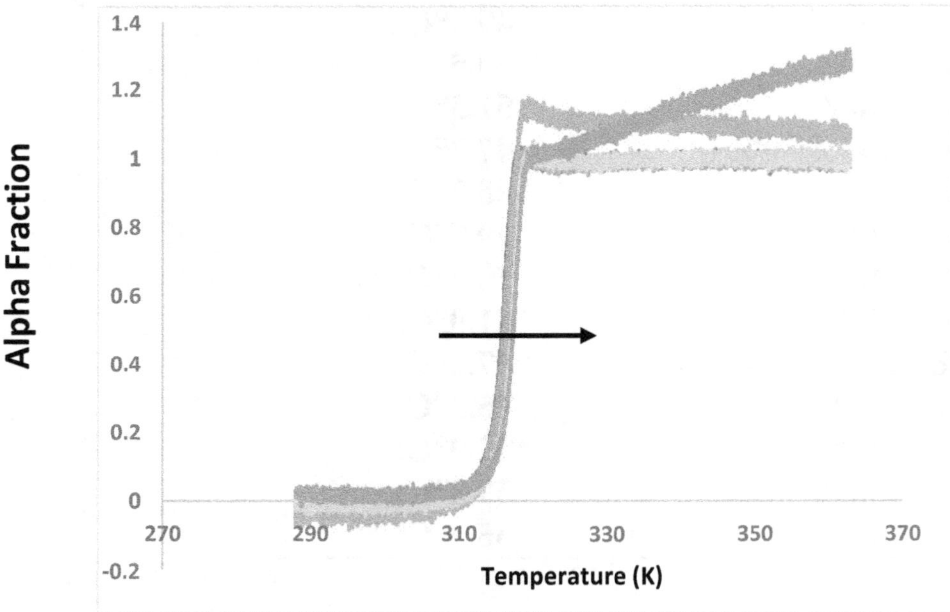

Figure 3.4.8: With increasing glycerol concentration, the melting temperature ofSpCas9 increases from 0% to 25% glycerol. Buffer Condition [Hepes - pH7.5(20mM),kCl(100mM),$MgCl_2$(10mM),DTT(1mM)]

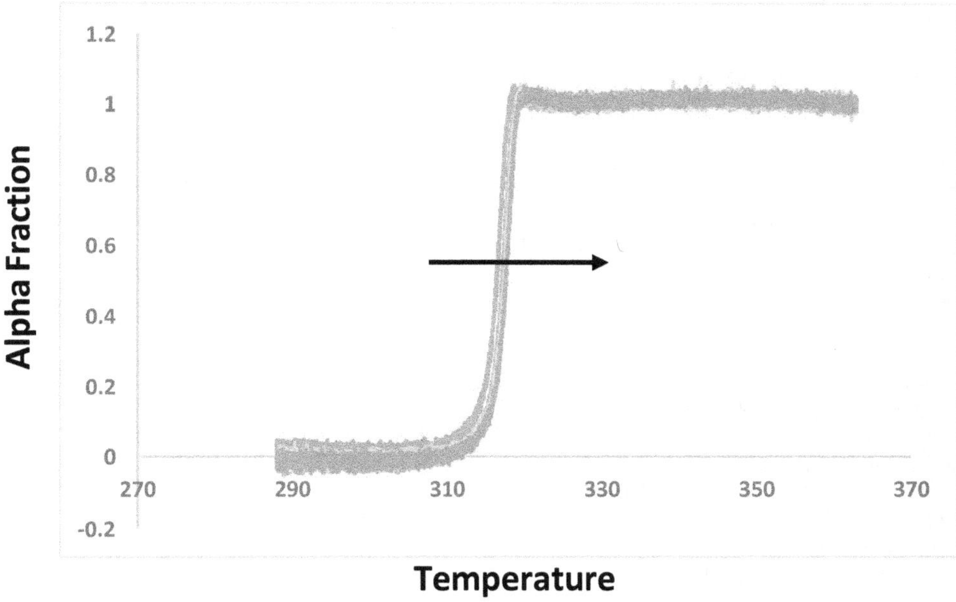

Figure 3.4.9: With increasing glycerol concentration, the melting temperature ofSpCas9 increases from 0% to 25% glycerol. Buffer Condition [Hepes- pH7.5(20mM),kCl(150mM),$MgCl_2$(10mM),DTT(1mM)]

Table 3.4.1. c: Thermodynamic properties calculated from Tm of WTSpCas9 by varying saltconcentration and osmolytes concentration:

	Series of experiments	Tm (Melting temp) (K)	ΔH (slope*R)(Kcal /mole)	ΔS (intercept*R)(Kcal/mole)	ΔG(kcal/mole) at room temp
1	SpCas9(0.4 mg/ml)(Hepes pH7.5(20mM),kCl(50mM),MgCl2(10mM),DTT(1 mM),glycerol(5%)	314.878	-199.548	-0.63372	-10.69944
2	SpCas9(0.4 mg/ml)(Hepes pH7.5(20mM),kCl(50mM),MgCl2(10mM),DTT(1 mM),glycerol(10%)	315.3681755	-244.862	-0.7767	-13.4054
3	SpCas9(0.4 mg/ml)(Hepes pH7.5(20mM),kCl(50mM),MgCl2(10mM),DTT(1 mM),glycerol(15%)	315.8612061	-231.67	-0.73372	-13.02144
4	SpCas9(0.4 mg/ml)(Hepes pH7.5(20mM),kCl(50mM),MgCl2(10mM),DTT(1 mM),glycerol(20%)	316.4906693	-342.974	-1.08374	-20.01948
5	SpCas9(0.4 mg/ml)(Hepes pH7.5(20mM),kCl(50mM),MgCl2(10mM),DTT(1 mM),glycerol(25%)	316.8418503	-304.116	-0.96002	-18.03004
6	SpCas9(0.4 mg/ml)(Hepes pH7.5(20mM),kCl(100mM),MgCl2(10mM),DTT(1 mM),glycerol(5%)	315.9604149	-185.072	-0.58562	-10.55724
7	SpCas9(0.4 mg/ml)(Hepes pH7.5(20mM),kCl(100mM),MgCl2(10mM),DTT(1 mM),glycerol(10%)	316.1661339	-253.414	-0.80182	-14.47164
8	SpCas9(0.4 mg/ml)(Hepes pH7.5(20mM),kCl(100mM),MgCl2(10mM),DTT(1 mM),glycerol(15%)	316.692894	-304.496	-0.96172	-17.90344
9	SpCas9(0.4 mg/ml)(Hepes pH7.5(20mM),kCl(100mM),MgCl2(10mM),DTT(1 mM),glycerol(20%)	317.1892319	-278.318	-0.8777	-16.7634
10	SpCas9(0.4 mg/ml)(Hepes pH7.5(20mM),kCl(100mM),MgCl2(10mM),DTT(1 mM),glycerol(25%)	317.4305191	-246.61	-0.77708	-15.04016
11	SpCas9(0.4 mg/ml)(Hepes pH7.5(20mM),kCl(150mM),MgCl2(10mM),DTT(1 mM),glycerol(5%)	315.9594307	-264.98	-0.83888	-14.99376
12	SpCas9(0.4 mg/ml)(Hepes pH7.5(20mM),kCl(150mM),MgCl2(10mM),DTT(1 mM),glycerol(10%)	316.4543533	-282.272	-0.89228	-16.37256
13	SpCas9(0.4 mg/ml)(Hepes pH7.5(20mM),kCl(150mM),MgCl2(10mM),DTT(1 mM),glycerol(15%)	317.1842422	-313.356	-0.98832	-18.83664
14	SpCas9(0.4 mg/ml)(Hepes pH7.5(20mM),kCl(150mM),MgCl2(10mM),DTT(1 mM),glycerol(20%)	317.4557304	-283.288	-0.8929	-17.2038
15	SpCas9(0.4 mg/ml)(Hepes pH7.5(20mM),kCl(150mM),MgCl2(10mM),DTT(1 mM),glycerol(25%)	317.640316	-225.288	-0.7093	-13.9166

Table3.4: Oligonucleotides used in the study

	DNA sequences used in the study	Oligonucleotide Sequence (5'-3')	Length
Primers for IVT template synthesis	VEGFA3_sgRNA _SpCa9_F.P	TAATACGACTCACTATAGGTGAGTGAGTGTG TGCGTGGTTTTAGAGCTAGAA	52
	Universal_SpCas 9_RP	AAAAGCACCGACTCGGTGCCACTTTTTCAAG TTGATAACGGACTAGCCTTATTTTAACTTGC TATTTCTAGCTCTAAAAC	80
Primers for MST experiment	FP_VEGFA3	TGTGGGTGAGTGAGTGTGTGCGTGTGGGGT	30
	RP_VEGFA3	ACCCCACACGCACACACTCACTCACCCACA	30
Primers for target DNA PCR product	VEGF3_T_500 FP	CTGGACACTTCCCAAAGGAC	20
	VEGF3_T_500_ RP	CTGAGAGCCGTTCCCTCTTT	20

3.4. Summary and Future Perspective

The efficiency of CRISPR-Cas9 genome-editing tool to cleave a target DNA is determinedspecifically by the components involved in the process of genome editing. The target DNAinterrogation is a consequence of three distinct events involving sustenance of Cas9 structure in ambient condition, structural changes of Cas9 upon sgRNA binding and finally, interaction of Cas9-sgRNA complex with target DNA substrate. Here, we showed the effect of salt and temperature in modulating all the above-mentioned steps. First, we aimedat observing the solution state structure and melting temperature of Cas9 and found that Cas9 has 66% alpha helical structure and 5% beta pleated sheet with a melting temperatureof 43.0(±0.5) °C in a two state fashion. The effect of salt perturbation on Cas9's structure and stability revealed thermal stability reaches to the highest with 500 mM salt concentration not showing significant change in the secondary structure. We also reportedthe quantitative physical properties encompassing the multi-step processes responsible for Cas9-based genome editing. Temperature-dependent binding studies of Cas9 with sgRNA interestingly reveals the interaction as exothermic process with negative entropy indicating the non-covalent bond formation in concomitant with Cas9 structural changes frominactive disordered to active ordered state which is in agreement with earlier report. On thecontrary, the binding of Cas9 RNP to target DNA is enthalpically unfavorable, entropicallyfavorable with mainly non electrostatic contribution to the overall energy. Of note, we observed relatively weak sgRNA binding to Cas9 (higher K_d) than that of Cas9 RNP withtarget DNA (lower K_d). Binding studies between Cas9 and sgRNA as well as Cas9 RNP and DNA resulted in favorable energy towards major non-electrostatic contribution overthe electrostatic counterpart when studied with varied salt concentration.

Taken together, the thermodynamic insights obtained from this study can be helpful in broadening the understanding of the CRISPR-Cas9 system which is essential for the development of a better tool for genome editing. The untended off-targeting ability of SpCas9-based system raises concerns about its use in therapeutic perspective to edit genomic DNA in biological system. Furthermore, this study can be extended to interpret the pattern of reaction energetics when CRISPR-Cas9 interrogates DNA off-targets. Thereis an immense scope to execute the bunch of assays used in this study, for off-target DNA to gather thermodynamic information like gibbs' free energy, enthalpy and entropy of such interaction. We may envision the thermodynamic parameters obtained from bindingstudies of RNP complex-DNA may give a drive towards the relatively unexplored field ofenergetics-based sgRNA designing tool and off-targeting evaluation platform. Considering the enormous off-targeting potential coming from widely used SpCas9 based systems, energetics-based platforms may give a new boost towards developing near perfect genomicscissor.

Alongside, we anticipate that the thermodynamic insights can also be extended to orthologous Cas9 systems owing to high conservation in Cas9's mode of action.

References:

1. Sorek, R., Kunin, V. & Hugenholtz, P. CRISPR — a widespread system that provides acquired resistance against phages in bacteria and archaea. *Nat. Rev. Microbiol.* **6**, 181–186 (2008).

2. Jinek, M. *et al.* A programmable dual-RNA-guided DNA endonuclease in adaptivebacterial immunity. *Science* **337**, 816–821 (2012).

3. Cong, L. *et al.* Multiplex genome engineering using CRISPR/Cas systems. *Science* **339**, 819–823 (2013).

4. Dai, W.-J. *et al.* CRISPR-Cas9 for in vivo Gene Therapy: Promise and Hurdles. *Mol. Ther. Nucleic Acids* **5**, e349 (2016).

5. Acharya, S., Maiti, S. & Chakraborty, D. Chapter 9 - CRISPR-Cas9 for therapy: thechallenges and ways to overcome them. in (eds. Singh, V. & Dhar, P. K. B. T.-G. E. via C.-C. S.) 101–110 (Academic Press, 2020). doi:https://doi.org/10.1016/B978-0-12-818140-9.00009-X.

6. Wang, D., Zhang, F. & Gao, G. CRISPR-Based Therapeutic Genome Editing: Strategies and In Vivo Delivery by AAV Vectors. *Cell* **181**, 136–150 (2020).

Pham, H., Kearns, N. A. & Maehr, R. Transcriptional Regulation with CRISPR/Cas9 Effectors in Mammalian Cells. *Methods Mol. Biol.* **1358**, 43–57 (2016).

7. Maeder, M. L. *et al.* CRISPR RNA-guided activation of endogenous human genes.

Nat. Methods **10**, 977–979 (2013).

8. Qi, L. S. Repurposing CRISPR as an RNA-guided platform for sequence-specificcontrol of gene expression. *Cell* **152**, (2013).

9. Chen, B. *et al.* Dynamic imaging of genomic loci in living human cells by an optimized CRISPR/Cas system. *Cell* **155**, 1479–1491 (2013).

10. Wang, H. *et al.* CRISPR-mediated live imaging of genome editing and transcription. *Science (80-.).* **365**, 1301 LP-- 1305 (2019).

11. Deng, W., Shi, X., Tjian, R., Lionnet, T. & Singer, R. H. CASFISH: CRISPR/Cas9-mediated in situ labeling of genomic loci in fixed cells. *Proc. Natl. Acad. Sci.* **112**, 11870 LP-- 11875 (2015).

12. Jusiak, B., Cleto, S., Perez-Piñera, P. & Lu, T. K. Engineering Synthetic Gene

Circuits in Living Cells with CRISPR Technology. *Trends Biotechnol.* **34**, 535–547(2016).

13. Gootenberg, J. S. *et al.* Nucleic acid detection with CRISPR-Cas13a/C2c2. *Science*
356, 438–442 (2017).
14. Chen, J. S. *et al.* CRISPR-Cas12a target binding unleashes indiscriminate single-stranded DNase activity. *Science* **360**, 436–439 (2018).
15. Azhar, M. *et al.* Rapid, field-deployable nucleobase detection and identification using FnCas9. *bioRxiv* 2020.04.07.028167 (2020) doi:10.1101/2020.04.07.028167.
16. Jiang, F. *et al.* Structures of a CRISPR-Cas9 R-loop complex primed for DNA cleavage. *Science* **351**, 867–871 (2016).
17. Jiang, F., Zhou, K., Ma, L., Gressel, S. & Doudna, J. A. STRUCTURAL BIOLOGY. A Cas9-guide RNA complex preorganized for target DNA recognition.*Science* **348**, 1477–1481 (2015).
18. Nishimasu, H. *et al.* Crystal structure of Cas9 in complex with guide RNA and target DNA. *Cell* **156**, 935–949 (2014).
19. Zhu, X. *et al.* Cryo-EM structures reveal coordinated domain motions that governDNA cleavage by Cas9. *Nat. Struct. Mol. Biol.* **26**, 679–685 (2019).

20. Huai, C. *et al.* Structural insights into DNA cleavage activation of CRISPR-Cas9system. *Nat. Commun.* **8**, 1375 (2017).

21. Dagdas, Y. S., Chen, J. S., Sternberg, S. H., Doudna, J. A. & Yildiz, A. A conformational checkpoint between DNA binding and cleavage by CRISPR-Cas9.*Sci. Adv.* **3**, eaao0027 (2017).

22. Yang, M. *et al.* The Conformational Dynamics of Cas9 Governing DNA Cleavage Are Revealed by Single-Molecule FRET. *Cell Rep.* **22**, 372–382 (2018).

23. Raper, A. T., Stephenson, A. A. & Suo, Z. Functional Insights Revealed by the Kinetic Mechanism of CRISPR/Cas9. *J. Am. Chem. Soc.* **140**, 2971–2984 (2018).

24. Sternberg, S. H. *et al.* HHHHS Public Access Author manuscript Nature. Author manuscript; available in PMC 2014 September 06. Published in final edited form as: Nature. 2014 March 6; 507(7490): 62–67. doi:10.1038/nature13011. DNA interrogation by the CRISPR RNA-guided endonucleas. *Nature* **507**, 62–67 (2014).

25. Jiang, F. & Doudna, J. A. CRISPR – Cas9 Structures and Mechanisms. 505–531(2017).

26. Xiang, G., Zhang, X., An, C., Cheng, C. & Wang, H. Temperature effect on CRISPR-Cas9 mediated genome editing. *J. Genet. Genomics* **44**, 199–205 (2017).

27. Dong, D. *et al.* The crystal structure of Cpf1 in complex with CRISPR RNA. *Nature* **532**, 522–526 (2016).

28. Wienken, C. J., Baaske, P., Rothbauer, U., Braun, D. & Duhr, S. Protein-binding assays in biological liquids using microscale thermophoresis. *Nat. Commun.* **1**, (2010).

29. Acharya, S. *et al.* Francisella novicida Cas9 interrogates genomic DNA with very high specificity and can be used for mammalian genome editing. *Proc. Natl. Acad.Sci. U. S. A.* **116**, 20959–20968 (2019).

30. Arora, A., Kaur, H., Wengel, J. & Maiti, S. Effect of locked nucleic acid (LNA) modification on hybridization kinetics of DNA duplex. *Nucleic Acids Symp. Ser.(Oxf)*. 417–418 (2008) doi:10.1093/nass/nrn212.

31. Yang, J. T., Wu, C. S. C. & Martinez, H. M. Calculation of Protein Conformation from Circular Dichroism. *Methods Enzymol.* **130**, 208–269 (1986).

32. Berman, H. M. *et al.* The protein data bank. *Acta Crystallogr. Sect. D Biol. Crystallogr.* **58**, 899–907 (2002).

33. Knapp, S., De Vos, W. M., Rice, D. & Ladenstein, R. Crystal structure of glutamatedehydrogenase from the hyperthermophilic eubacterium Thermotoga maritima at 3.0 Å resolution. *J. Mol. Biol.* **267**, 916–932 (1997).

34. Richter, F. *et al.* Engineering of temperature- and light-switchable Cas9 variants.

Nucleic Acids Res. **44**, 10003–10014 (2016).

35. Reviews, Q., Cells, G. & Chemistry, B. QRB_1997_30_241_Cacace. 1–37 (1997).

36. Zhang, Y. & Cremer, P. S. Interactions between macromolecules and ions: the Hofmeister series. *Curr. Opin. Chem. Biol.* **10**, 658–663 (2006).

37. Tadeo, X., López-Méndez, B., Castaño, D., Trigueros, T. & Millet, O. Protein stabilization and the hofmeister effect: The role of hydrophobic solvation. *Biophys.J.* **97**, 2595–2603 (2009).

38. Strickler, S. S. *et al.* Protein stability and surface electrostatics: A charged relationship. *Biochemistry* **45**, 2761–2766 (2006).

39. Meeker, A. K., Garcia-Moreno E., B. & Shortle, D. Contributions of the ionizable amino acids to the stability of staphylococcal nuclease.

Biochemistry **35**, 6443–6449(1996).

40. Record, M. T., Anderson, C. F. & Lohman, T. M. Thermodynamic analysis of ioneffects on the binding and conformational equilibria of proteins and nucleic acids:The roles of ion association or release, screening, and ion effects on water activity.*Q. Rev. Biophys.* **11**, 103–178 (1978).

41. Wang, H., La Russa, M. & Qi, L. S. CRISPR/Cas9 in Genome Editing and Beyond.

Annu. Rev. Biochem. **85**, 227–264 (2016).

42. Hsu, P. D. *et al.* DNA targeting specificity of RNA-guided Cas9 nucleases. *Nat. Biotechnol.* **31**, 827–832 (2013).

43. Yamada, M. *et al.* Crystal Structure of the Minimal Cas9 from Campylobacter jejuni Reveals the Molecular Diversity in the CRISPR-Cas9 Systems. *Mol. Cell* **65**,1109-1121.e3 (2017).

44. Nishimasu, H. *et al.* Crystal structure of Cas9 in complex with guide RNA and target DNA. *Cell* **156**, 935–949 (2014).

45. Nishimasu, H. *et al.* Crystal Structure of Staphylococcus aureus Cas9. *Cell* **162**, 1113–1126 (2015)

CSIR 800 Project Summary

Genetics and genetic testing of Congenital myasthenia Syndrome

CONTENT

This information booklet was created as part of the **CSIR-800** programme of AcSIR by **Ms. Dipanjali Sinha** in partial fulfillment of the requirements for award of the PhDdegree.

CSIR-800

The CSIR-800 program aims for Inclusive growth and improved quality of life for 800million citizens at the base of the Indian economic pyramid, through innovative scienceand technology interventions that are socially and economically relevant. For more information: **http://www.csir800.org/.**

The AcSIR CSIR-800 Programme

The objective of the Academy of Scientific and Innovative Research (AcSIR) is to disseminate advanced knowledge in science and technology, particularly in emerging interdisciplinary and multi-disciplinary areas to create socially conscious, highest quality personnel. The ultimate aim is to create human resources who will promote research in science and technology having a bearing on social economic, cultural, intellectual and academic welfare of the people of our nation. Keeping this in view AcSIR has mandated that students aspiring to obtain a Ph.D. degree from the academy undertake a 6-8 weeks project concerned with societal/ rural issues under the CSIR-800Programs (4 credits). CSIR-800 has been launched by CSIR with the aspiration of inclusive growth and improved quality of lives of 800 million fellow Indians through S&T interventions that are socially and economically relevant.

Financial Disclosure:

The author acknowledges funding and support from the Council of Scientific and Industrial Research (CSIR), India through Grant No BSC0122 (CARDIOMED).

What is Congenital Myasthenia Syndrome?

Congenital Myasthenia (CM) syndrome is a rare genetic disease characterized by muscle weakness (myasthenia) that worsens with physical strain. The muscle weakness generally begins in early childhood but can appear in adolescence or adulthood. Facial muscles, including muscles that control the eyelids and the movement of the eyes, and muscles used for chewing and swallowing are most commonly affected. Any of the muscles used for movement like skeletal muscles can be affected in this condition. Development of motor skills like crawling or walking may be delayed in the affected. The severity of the myasthenia varies greatly with some people experiencing minor weakness and others having such severe weakness that they are not able to walk. CHRNE gene encodes a protein acetylcholine receptor subunit epsilon in humans. At mature mammalian neuromuscular junctions, acetylcholine receptors are pentameric protein complexes composed of four subunits in the ratio of two alpha subunits to one beta, one epsilon, and one delta subunit. The acetylcholine receptor changes subunit composition shortly after birth when the epsilon subunit replaces the gamma subunit found in embryonic receptors. Causing the mutations in the epsilon subunit are associated with congenital myasthenic syndrome.

Figure 1: Chromosomal location of CHRNE gene (NCBI).

How common is CMS Syndrome?

Congenital Myasthenia syndrome is a rare genetic disease with only a few reported cases. The mutation leading to the disease has low penetrance i.e. not all of the people with the mutation show disease phenotype. About 40% of the carriers (heterozygous individuals) show disease phenotype by the age of 40.

What are the symptoms found in CMS Syndrome patients?

Congenital myasthenic syndrome is a group of conditions characterized by muscle weakness that worsens with physical strain. The muscle weakness begins in childhoodbut can appear in adolescence or adulthood. CMS syndrome is a disease that can manifest as early as ten years of age. In some patients with CMS, the weakness is confined to muscles supplied by the cranial nerves causing double vison, droopy eyelids (eyelid ptosis), facial weakness, hypernasal or slurred speech, and swallowing difficulties. Congenital myasthenia affects the voluntary muscles of the body, especially those that control the eyes, mouth, throat and limbs. The disease can strike anyone at any age, but is more frequently seen in young women of age 20 and 30 and men aged 50 and older.

What is the function of CHRNE gene?

The CHRNE gene instructs for making the epsilon (ε) component /subunit of the acetylcholine receptor (AChR) protein. Acetylcholine receptors are pentameric protein complexes composed of four subunits in the ratio of two alpha subunits to one beta, one epsilon, and one delta subunit at mature mammalian neuromuscular junctions. The acetylcholine receptor changes subunit composition shortly after birth while the epsilonsubunit replaces the gamma subunit in embryonic stage. Now, mutations in the epsilonsubunit are associated with congenital myasthenic syndrome.

The CHRNE gene encodes for the epsilon subunit of the AChR. Most mutations are autosomal recessive and of loss-of-function happens and as a result there is endplate AChR deficiency. CHRNE gene is associated with changing the kinetic properties of the AChR. One type of mutation of the epsilon subunit of the AChR occurs via introducing an arginine (Arg) into the binding site at the α/ε subunit interface of the receptor. The addition of a cationic Arg into the anionic environment of the AChR binding site greatly reduces the kinetic properties of the receptor while the mutation is introduced. The consequence of the newly introduced Arg to the epsilon subunit, is a 30-fold reduction of agonist affinity, 75-fold reduction of gating efficiency, and an extremely weakened channel opening probability. Mutation like this results in an extremely fatal form of CMS.

How does mutations cause CMS syndrome?

This CMS is caused by dominant mutations in the ligand binding or pore domains of AchR resulting in prolonged synaptic currents and potentials. Mutation in the ligand binding domain enhances affinity for ACh by slowing its dissociation rate from the doubly-liganded receptor. This mechanism prolongs the single-channel currents and EPcurrents while mutations in the pore domain cause spontaneous channel openings. In EP AChE deficiency, the prolonged EPP triggers a repetitive CMAP, but in this syndrome edrophonium enhances the repetitive response in patient. During activity, prolonged EPPs progressively depolarize the postsynaptic membrane causing a permanent depolarization block. The prolonged and spontaneous openings of the AChRchannel results in Ca^{2+} accumulation in the postsynaptic region and an EP myopathy.

How do people inherit CMS syndrome?

CMS syndrome is autosomal recessive hereditary disease which means it could be passing down through families. The individual who gets one diseased allele from any of the parent will be the carrier. Carriers are usually unaffected, but can pass the faulty allele onto their off-springs. If one or both the parents is/are carriers, there is 50% possibility that each of their offspring can be a carrier. The offspring who gets faulty alleles from both parents is more likely to have the disease as there is 25% risk that theoffspring is affected. Thus, CMS syndrome occurs only when both parents carry and transmits their mutant allele to the child.

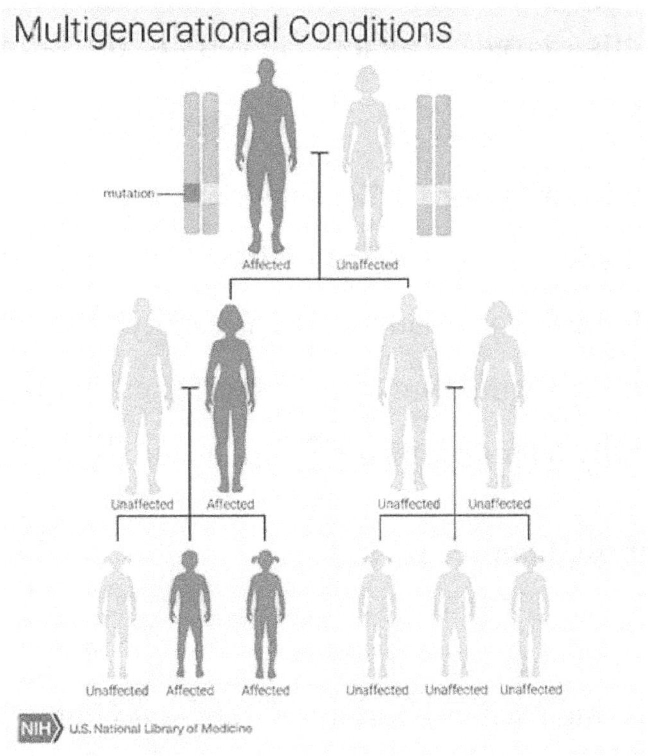

Figure 2: Autosomal recessive hereditary pattern of CMS syndrome disease

Are the exact mutations which cause the disease known?

According to the NCBI ClinVar (http://www.ncbi.nlm.nih.gov/clinvar/), there are 42 pathogenic mutations in CHRNE gene reported till date. These consist of base pair insertions, deletions, duplications and point mutations. The protein product is altered by all of the mutations. As a consequence of these mutations, the protein would be non-functional resulting from a defect at the junction where your nerve stimulates muscle activity.

How it can be diagnosed? Can any carriers in the family be diagnosed?

There is prenatal testing for CMS mutations available under various panels such as the Pediatric Cancer panel (Prevention Genetics), WES Muscular disorders (Translation Medical Lab, Radbound University) and multiple panels from Ambry Genetics. Individuals who want to get tested for carrier mutations, can do so by opting for one ofthe many testing companies and hospitals available.

Can it be diagnosed before birth?

Yes, prenatal testing such as chorionic villus sampling (CVS) and amniocentesis, can diagnose CMS syndrome. CVS can be performed almost in the 11th week of pregnancywhich includes removing a tiny piece of placenta. The alternative way is amniocentesis,which is performed in almost 16th month of pregnancy. The fluid surrounding the baby is removed using a needle and will be subjected to test. Genetic testing can also be donefor the parents to find if either or both of them are carriers.

Why is genetic diagnosis important?

Genetic diagnosis is necessary as it helps to identify the variant and thereby confirm whether we have the risk allele or not. This could confirm a proper clinical test which would make a better picture about the condition. This can also suggest us to screen other members in the family for carrier status and could remind us to perform prenatal diagnosis on fetal samples or even track the genetic inheritance.

Genetic diagnosis can help recommending a genetic counseling for risks on having other children and guides in estimating how the disease might progress.

Is there a cure? Treatment for CMS syndrome ?

There is no absolute cure for CMS syndrome as there are usually no straightforward treatments to cure the underlying genetic abnormality. The prognosis depends on the specific subtype of congenital myasthenia, the specific muscles involved, and the age at onset of symptoms. The patients with the disease or the causal mutation should exercise and eat healthy.

Recent therapies for CMS involve medications which is cholinergic agonists, amifampridine i.e. (3,4-diaminopyridine) or pyridostigmine, open channel blockers of acetylcholine receptor ion channel which is also long lived like quinidine, fluoxetine and adrenergic agonists which are salbutamol and ephedrine.

Is there any prevention for the disease?

As there is no prevention for Congenital Myasthenia syndrome, it is highly recommended to have a genetic testing before getting married if the couple is consanguineous. In pregnancy stage, testing of amniotic fluid around fetus could also diagnose the disease.

Where can I go for genetic testing?

Currently DNA Labs and CMC Vellore in India are performing the genetic testing for CMS syndrome as part of their panels. However, specific testing of the gene can be performed by sending samples to labs outside India where there are more diseases screened at one go.

References for CSIR 800:

1. Andrew G. Engel, Xin-Ming Shen, Duygu Selcen, and Steven M. Sine. Congenitalmyasthenic syndromes: pathogenesis, diagnosis, and treatment. *Lancet Neurol*. 2015April ; 14(4): 420–434. doi:10.1016/S1474-4422(14)70201-7

2. Josef Finsterer . Congenital myasthenic syndromes . Finsterer Orphanet Journal ofRare Diseases[https://doi.org/10.1186/s13023-019-1025-5]

3. https://rarediseases.org/rare-diseases/congenital-myasthenic-syndromes/

4. Angela Abicht, MD, Juliane S Müller, PhD, and Hanns Lochmüller, MD.,Congenital Myasthenic Syndromes Overview

5. https://www.mayoclinic.org/diseases-conditions/congenital-myasthenic-syndrome/symptoms-causes/syc-20354754

https://practicalneurology.com/articles/2019-aug-july/congenital-myasthenic-syndrome-treatment

6. https://en.wikipedia.org/wiki/CHRNE

7. https://www.ncbi.nlm.nih.gov/gene/1145

www.ingramcontent.com/pod-product-compliance
Lightning Source LLC
LaVergne TN
LVHW082354200926
848481LV00012B/3563
9798230131236